Power Practice Problems

for the Electrical and Computer PE Exam

John A. Camara, PE

The Power to Pass®
www.ppi2pass.com

Professional Publications, Inc. • Belmont, California

Benefit by Registering This Book with PPI

- Get book updates and corrections.
- Hear the latest exam news.
- Obtain exclusive exam tips and strategies.
- Receive special discounts.

Register your book at **www.ppi2pass.com/register**.

Report Errors and View Corrections for This Book

PPI is grateful to every reader who notifies us of a possible error. Your feedback allows us to improve the quality and accuracy of our products. You can report errata and view corrections at **www.ppi2pass.com/errata**.

NFPA 70®, *National Electrical Code*®, and NEC® are registered trademarks of the National Fire Protection Association, Inc., Quincy, MA 02169.

National Electrical Safety Code® and NESC® are registered trademarks of the Institute of Electrical and Electronics Engineers, Inc., New York, NY 10016.

POWER PRACTICE PROBLEMS FOR THE ELECTRICAL AND COMPUTER PE EXAM

Current printing of this edition: 2

Printing History

edition number	printing number	update
1	1	New book.
1	2	Minor corrections.

PPI
1250 Fifth Avenue, Belmont, CA 94002
(650) 593-9119
www.ppi2pass.com

ISBN: 978-1-59126-164-3

Library of Congress Control Number: 2010931172

Topics

Where do I find help solving these Practice Problems?

Power Practice Problems for the Electrical and Computer PE Exam presents complete, step-by-step solutions for 443 problems covering topics on the Power exam. This book is a companion to the *Power Reference Manual*; the subjects covered in the two books correspond chapter for chapter, and the solutions in this book often cite the *Reference Manual*. You can find all the background information you need to solve these problems—including equations, figures, charts, and tables of data—in the *Power Reference Manual*.

The *Power Reference Manual* may be purchased directly from PPI at **www.ppi2pass.com** or from your favorite bookstore.

Table of Contents

The topics and chapters listed here correspond to those in the *Power Reference Manual*.

Topic XII: Measurement and Instrumentation

Topic XIII: Electrical Materials

Topic XIV: Codes and Standards

Topic XV: Professional

Preface

Power Practice Problems for the Electrical and Computer PE Exam is the companion book to the *Power Reference Manual for the Electrical and Computer PE Exam*. *Power Practice Problems* contains problems on a myriad of topics, encompassing the fundamentals of the electrical engineering field, including topics appearing on the NCEES Power exam.

As you work the problems presented here, you will prepare for the Power exam in the most efficient way possible. The problems are written to suggest areas of further study; should you have difficulty with a topic, further explanation may be found in the corresponding chapter of the *Power Reference Manual*. Incorrect answers enlighten you as to common mistakes, and the steps necessary to arrive at correct answers are thoroughly documented. When you take full advantage of the practice problems and solutions, you will gain knowledge that is essential for peak performance on your exam.

Power Practice Problems descended from the eighth edition of *Electrical Engineering Practice Problems for the Power, Electrical and Electronics, and Computer PE Exams*. The first edition of *Electrical Engineering Practice Problems* was published in 1978. Over its many years in print, it has evolved with the industry and with changes to the Electrical and Computer PE exam. In April 2009, NCEES made a fundamental shift in the focus of this exam, converting the single morning session and three specialized afternoon sessions into three independent full-day specialized exams—for Power, for Electrical and Electronics, and for Computer Engineering. I have written *Power Practice Problems* to expand and focus the power coverage of *Electrical Engineering Practice Problems* into a text supplying what you need to succeed on the Power exam.

WHAT'S IN THIS BOOK

Along with revising sections from the best-selling aforementioned practice problems book, I added new chapters on power system analysis, protection and safety, and power system management. I also expanded the chapters on rotating DC machinery, rotating AC machinery, and illumination.

While I wrote this book with the NCEES specifications in mind, the contents of *Power Practice Problems* are not limited to the specifications of the Power exam. This reference also contains practice problems on the fundamental blocks you need to build a complete understanding of power engineering. With an understanding of the basic problems that arise in electrical engineering, you will comprehend the majority of exam questions, as well as real-world electrical engineering problems.

The sources I consulted while developing this book's content include: (1) information on the Power exam published by NCEES; (2) *PE Electrical and Computer: Power Sample Questions and Solutions*, published by NCEES; (3) electrical engineering curricula at leading colleges and universities; (4) current literature in the field of electrical engineering; (5) electrical engineering websites; and (6) survey comments from those who have recently taken the PE exam and/or who purchased *Electrical Engineering Practice Problems*.

A number of features enhance the usefulness of this text. Each problem contains a full solution, explaining the correct answer as well as the incorrect choices. Lists of codes and references recommended for additional study are supplied. The publisher has also provided an excellent resource in support of your success: you may access online updates, report errors, and participate in discussions with other prospective electrical engineers through PPI's website, **www.ppi2pass.com**.

FOR WHOM THIS BOOK IS WRITTEN

Power Practice Problems is written for you. If you are an exam candidate, *Power Practice Problems* is an efficient resource for applying exam topics systematically and exhaustively. If you are a practicing power engineer, it functions as a companion to the *Power Reference Manual*, expanding knowledge of all aspects of this field in a realistic manner, and incorporating the most common formulas and data. Finally, for the engineering student, it presents a thorough review of the fundamentals of electrical engineering.

Future editions of this book will be very much shaped by what you and others want to see in it. I am braced for the influx of comments and suggestions from readers who (1) want more problems in some chapters, (2) want more detail in existing problems, and (3) think that continuing to include some problems is just plain lame. Computer hardware and programming change quickly, as does the grand unified (atomic) theory of everything. I expect several chapters to generate some kind of debate.

Should you find an error in this book, know that it is mine, and that I regret it. Beyond that, I hope two things happen. First, please let me know about the error by using the error reporting form on the PPI website, found at **www.ppi2pass.com/errata**. Second, I hope you learn something from the error—I know I will! I appreciate suggestions for improvement, additional questions, and recommendations for expansion so that new editions or similar texts will better meet future examinees' needs.

Read carefully, prepare well, and you will triumph over the exam!

John A. Camara, PE

Acknowledgments

It is with enduring gratitude that I thank Michael R. Lindeburg, PE, who, with the sixth editions of the *Electrical Engineering Reference Manual* and *Electrical Engineering Practice Problems*, allowed me to realize my dream of authoring significant engineering texts. In taking the reins from the titles' previous author, the late Raymond Yarbrough, I was entrusted with the responsibility of shaping those texts as the electrical and computer PE exam and the engineering profession evolved. It is my hope that *Power Practice Problems for the Electrical and Computer PE Exam*—which is based on *Electrical Engineering Practice Problems*—will match the comprehensive and cohesive quality of the best of PPI's family of engineering books, which I have admired for more than three decades.

Power Practice Problems would not have been realized without the professionalism and meticulous attention to detail shown by Megan Synnestvedt, copy editor. Without her spreadsheets, I would have lost track of chapter numbers right from the beginning! In addition, my thanks go out to the wonderfully responsive team at PPI, including Meaghan Banks, editorial project manager; Sarah Hubbard, director of new product development, whose guidance from the beginning has been most appreciated; Cathy Schrott, director of production, who keeps it all moving forward; copy editors Tyler Hayes, Jenny Lindeburg King, Scott Marley, Magnolia Molcan, Sesa Pabalan, and Trina Tom; acquisitions assistant Katie Throckmorton for obtaining permissions; Tom Bergstrom, technical illustrator; Alex Foster and Kate Hayes, the vitally important typesetters; and Amy Schwertman, who designed the cover.

Gregg Wagener, PE, reviewed each of the chapters of this book, providing valuable insight, correcting numerous errors and, in general, teaching me a great deal along the way.

Michael Lindeburg provided the original Chapters 1–14 on mathematics, which I have expanded, as well as the chapters on economic analysis, law, ethics, and engineering licensure. These contributions saved me considerable time, and I am grateful to use material of this quality. John Goularte, Jr., taught me first-hand the applications of biomedical engineering, and the real-world practicality of this topic's problems is attributable to him.

The knowledge in this book represents years of training, instruction, and self-study in electrical, electronic, mechanical, nuclear, marine, space, and a variety of other types of engineering—all of which I find fascinating. I would not understand any of it were it not for the teachers, instructors, mentors, family, and friends who've spent countless hours with me and from whom I've learned so much. A few of them include: Mrs. Mary Avila, Capt. C. E. Ellis, Mr. Claude Estes, Capt. Karl Hasslinger, Mr. Jerome Herbeck, Mr. Jack Hunnicutt, Mr. Ralph Loya, Mr. Harry Lynch, Mr. Harold Mackey, Mr. Michael O'Neal, Mr. Mark Richwine, Mr. Charles Taylor, Mrs. Abigail Thyarks, and Mr. Jim Triguerio. There are others who have touched my life in special ways: Jim, Marla, Lora, Todd, Tom, and Rick. There are many others, though here they will remain nameless, to whom I am indebted. Thanks.

Finally, thanks to my wife, Becky, and my mother, Arlene, both of whom have made me a better person. In addition to providing valuable content guidance, Becky was, more importantly, patient and supportive through the entire process, even when I wasn't.

John A. Camara, PE

Codes and References

Power Practice Problems for the Electrical and Computer PE Exam is based on the following codes and references. The electrical PE exam is consistent with the last edition of the code issued before the year of the exam. For more information about the variety of codes related to electrical engineering, refer to the following organizations and their websites.

American National Standards Institute (www.ansi.org)
Electronic Industries Alliance (www.eia.org)
Federal Communications Commission (www.fcc.gov)
Institute of Electrical and Electronic Engineers
   (www.ieee.org)
Instrument Society of America (www.isa.org)
International Organization for Standardization
   (www.iso.org)
National Electrical Manufacturer's Association
   (www.nema.org)
National Fire Protection Agency (www.nfpa.org)

The PPI website (**www.ppi2pass.com/eefaq**) provides the dates of the codes, standards, and regulations that the current exams are based on, according to NCEES. Use this information to decide which editions of these books should be part of your exam preparation.

The minimum recommended library of materials to bring to the electrical PE exams consists of this book, the applicable code book, a standard handbook of electrical engineering, and one or two textbooks that cover fundamental circuit theory (both electrical and electronic). The following list includes codes necessary for taking the exam, as well as textbooks and handbooks that you may be familiar with and that would be useful resources while taking and preparing for the exam. The books listed with "reference" brackets are those that should prove useful to you after the exam.

CODES

Federal Communications Commission (FCC). *Code of Federal Regulations*, "FCC Rules": 2004 Title 47-Telecommunications, Ch. 73. "Radio Broadcast Rules," (47CFR73). [Communications]

Institute of Electrical and Electronics Engineers (IEEE). 2008 *National Electrical Safety Code*, (NESC). [Power]

National Fire Protection Association (NFPA). 2008 *National Electrical Code*, (NFPA 70 or NEC). [Power]

REFERENCES

American Electrician's Handbook. Terrell Croft and Wilford I. Summers. McGraw-Hill. [Power handbook]

An Introduction to Digital and Analog Integrated Circuits and Applications. Sanjit K. Mitra. Harper & Row, Publishers. [Digital Circuit Fundamentals textbook]

Applied Electromagnetics. Martin A. Plonus. McGraw-Hill. [Electromagnetic Theory textbook]

CRC Materials Science and Engineering Handbook. James F. Shackelford, William Alexander, and Jun S. Park, eds. CRC Press, Inc. [General engineering handbook]

CRC Standard Mathematical Tables. William H. Beyer, ed. CRC Press, Inc. [General engineering reference]

Electrical Power Technology. Theodore Wildi. John Wiley & Sons. [Power Theory and Application textbook]

Linear Circuits. M. E. Van Valkenburg and B. K. Kinariwala. Prentice-Hall, Inc. [AC/DC Fundamentals textbook]

McGraw-Hill Dictionary of Scientific and Technical Terms. Sybil P. Parker, ed. McGraw-Hill. [General engineering reference]

Microelectric Circuit Design. Richard C. Jaeger. McGraw-Hill. [Electronic Fundamentals textbook]

Microelectronics. Jacob Millman. McGraw-Hill. [Electronic Fundamentals textbook]

National Electrical Code Handbook. Mark W. Earley, Joseph V. Sheehan, et al. National Fire Protection Association. [Power handbook]

National Electrical Safety Code (NESC) Handbook. David J. Marie. McGraw-Hill. [Power handbook]

NEC Answers. Michael A. Anthony. McGraw-Hill. [*National Electrical Code* Example Applications textbook]

Power System Analysis. John J. Grainger and William D. Stevenson, Jr. McGraw-Hill. [Power textbook]

Process/Industrial Instruments and Controls Handbook. Gregory K. McMillan, ed. McGraw-Hill. [Power and Electrical and Electronics handbook]

Standard Handbook for Electrical Engineers. Donald G. Fink and H. Wayne Beaty. McGraw-Hill. [Power and Electrical and Electronics handbook]

The Biomedical Engineering Handbook. Joseph D. Broizmo, ed. CRC Press. [Electrical and Electronics handbook]

The IESNA Lighting Handbook, Reference & Applications. Mark S. Rea, ed. Illuminating Engineering Society of North America. [Power handbook]

Wireless and Cellular Telecommunications. William C.Y. Lee. McGraw-Hill. [Electrical and Electronics handbook]

How to Use This Book

This book is a companion to the *Power Reference Manual*—a compendium of power engineering. Admittedly, despite its size, the *Reference Manual* cannot cover all the details of power engineering. Nevertheless, a thorough study of that volume and its equations, terms, diagrams, and tables, as well as the problem statements and solutions given in this *Practice Problems* book, will expose you to many electrical engineering topics. This broad overview of the field can benefit you in a number of ways, not the least of which is by thoroughly preparing you for the Power exam. Further, the range of subjects covered allows you to tailor your review toward the topics of your choice, and to study at a level that suits your needs.

Before you start studying, read about the format and content of the Power exam in the *Power Reference Manual* or on the PPI website at **www.ppi2pass.com/eefaqs**. Then, begin your studies with the fundamentals and progressively expand your review to include more specific topics. Keep in mind that experience in problem solving is key to success on the exam.

Depending on your specific study methods, you may wish to study the "easier" material first to build confidence. Alternatively, you might delay reading the easy material until just prior to the exam. The choice is yours.

If you are using this text to practice a particular area of study, go directly to the topic of interest and begin. Any weakness noted while attempting to solve the problems should prompt you to review the applicable material in the *Reference Manual*.

Allow yourself three to six months to prepare. You'll get the most benefit out of your exam preparation efforts if you make a plan and stick to it. Make sure, as you review chapters in the accompanying *Reference Manual*, that you also leave enough time to work the problems associated with those chapters. You should revisit practice problems during your exam preparation. Even if you successfully worked all the problems in a given chapter upon your initial review, you must maintain that knowledge as you work through unfamiliar or difficult material.

I'm confident that with diligent study, you will find the electrical engineering knowledge you seek within these pages or within the pages of the *Reference Manual*. At the very least, the companion book will allow you to conduct an intelligent search for knowledge. I hope these texts serve you well. Enjoy the adventure of learning!

John A. Camara, PE

1 Systems of Units

PRACTICE PROBLEMS

1. Convert 250°F to degrees Celsius.

(A) 115°C

(B) 121°C

(C) 124°C

(D) 420°C

2. Convert the Stefan-Boltzmann constant $(1.71 \times 10^{-9}$ Btu/ft²-hr-°R⁴) from customary U.S. to SI units.

(A) 5.14×10^{-10} W/m²·K⁴

(B) 0.950×10^{-8} W/m²·K⁴

(C) 5.67×10^{-8} W/m²·K⁴

(D) 7.33×10^{-6} W/m²·K⁴

3. How many U.S. tons (2000 lbm per ton) of coal with a heating value of 13,000 Btu/lbm must be burned to provide as much energy as a complete nuclear conversion of 1 g of its mass? (Hint: Use Einstein's equation: $E = mc^2$.)

(A) 1.7 tons

(B) 14 tons

(C) 780 tons

(D) 3300 tons

4. What is the SI unit for force, N, in terms of more basic units?

(A) kg·m/s²

(B) kg·m²/s²

(C) kg·m²/s³

(D) kg/m·s²

5. What is the appropriate SI unit for magnetic flux density, **B**?

(A) henry, H

(B) siemens, S

(C) tesla, T

(D) weber, W

SOLUTIONS

1. The conversion of temperature Fahrenheit (T_F) to Celsius (T_C) is

$$T_C = \left(\tfrac{5}{9}\right)(T_F - 32)$$
$$= \left(\tfrac{5}{9}\right)(250°F - 32°F)$$
$$= \left(\tfrac{5}{9}\right)(218°F)$$
$$= \boxed{121.1°C \quad (121°C)}$$

The answer is (B).

2. In customary U.S. units, the Stefan-Boltzmann constant, σ, is 1.71×10^{-9} Btu/ft²-hr-°R⁴.

Use the following conversion factors.

$$1 \text{ Btu/hr} = 0.2931 \text{ W}$$
$$1 \text{ ft} = 0.3048 \text{ m}$$
$$T_R = \tfrac{9}{5} T_K$$

Performing the conversion gives

$$\sigma = \left(1.71 \times 10^{-9} \frac{\text{Btu}}{\text{hr-ft}^2\text{-°R}^4}\right)\left(0.2931 \frac{\text{W}}{\frac{\text{Btu}}{\text{hr}}}\right)$$
$$\times \left(\frac{1 \text{ ft}}{0.3048 \text{ m}}\right)^2 \left(\frac{9}{5}\frac{°R}{K}\right)^4$$
$$= \boxed{5.67 \times 10^{-8} \text{ W/m}^2\text{·K}^4}$$

The answer is (C).

3. The energy produced from the nuclear conversion of any quantity of mass is given as

$$E = mc^2$$

The speed of light, c, is 2.9979×10^8 m/s.

For a mass of 1 g (0.001 kg),

$$E = mc^2$$
$$= (0.001 \text{ kg})\left(2.9979 \times 10^8 \frac{\text{m}}{\text{s}}\right)^2$$
$$= 8.9874 \times 10^{13} \text{ J}$$

Convert to U.S. customary units with the conversion 1 Btu = 1055 J.

$$E = (8.9874 \times 10^{13} \text{ J}) \left(\frac{1 \text{ Btu}}{1055 \text{ J}} \right)$$

$$= 8.5189 \times 10^{10} \text{ Btu}$$

The number, n, of tons of 13,000 Btu/lbm coal is

$$n \text{ tons} = \frac{8.5189 \times 10^{10} \text{ Btu}}{\left(13{,}000 \ \frac{\text{Btu}}{\text{lbm}} \right) \left(2000 \ \frac{\text{lbm}}{\text{ton}} \right)}$$

$$= \boxed{3276 \text{ tons} \quad (3300 \text{ tons})}$$

The answer is (D).

4. The newton, N, is the SI unit of force. Using primary dimensional analysis,

$$F = \frac{ML}{\theta^2}$$

$$= \boxed{\text{kg·m/s}^2}$$

The answer is (A).

5. The SI unit for magnetic flux density, **B**, is the tesla, T.

The answer is (C).

2 Energy, Work, and Power

PRACTICE PROBLEMS

1. A volume of 40×10^6 m³ of water behind a dam is at an average height of 100 m above the input to electrical turbines. The turbines are 90% efficient. What is the approximate maximum potential energy converted to electrical energy?

(A) 3.9×10^{11} J

(B) 3.5×10^{13} J

(C) 3.9×10^{13} J

(D) 1.3×10^{14} J

2. A certain motor parameter is measured by the following expression.

$$\int \mathbf{T} \cdot d\theta$$

What is the motor parameter being measured?

(A) power

(B) thrust

(C) torque

(D) work

3. A linear spring, held by an electric coil, is designed to open a cooling valve in the event of an electrical failure. The conditions are as shown.

de-energized   **energized**

What work must the electric coil do to keep the spring compressed?

(A) 0.018 N·m

(B) 0.075 N·m

(C) 750 N·m

(D) 75×10^3 N·m

4. Consider the illustration shown. An induction motor is rotating at 1800 rpm. The mass is concentrated on the circumference of the rotor. What is the approximate kinetic energy of the rotor?

$m = 100$ kg

0.3 m

(A) 2.8×10^3 J

(B) 1.6×10^5 J

(C) 3.2×10^5 J

(D) 6.4×10^5 J

5. A rotating motor is calculated to have 20×10^4 J of rotational energy during operation. During fault analysis, this motor is expected to dissipate all this energy in one second through a ground. What power is provided to the ground site by the motor?

(A) 20 W

(B) 200 W

(C) 200 kW

(D) 400 kW

SOLUTIONS

1. Using the average height of 100 m, the potential energy, $E_{available}$, is given by the product of the mass of the water (its density, ρ_{H_2O}, times its volume), the acceleration of gravity, g, and the height of the dam above the turbines, z.

$$E_{available} = mgz$$
$$= (\rho_{H_2O} V_{dam})gz$$
$$= \left(1000 \ \frac{kg}{m^3}\right)(40 \times 10^6 \ m^3)$$
$$\times \left(9.81 \ \frac{m}{s^2}\right)(100 \ m)$$
$$= 3.924 \times 10^{13} \ J$$

At an efficiency, η, of 90%, the maximum energy output is

$$E_{output} = \eta E_{available}$$
$$= (0.90)(3.924 \times 10^{13} \ J)$$
$$= \boxed{3.53 \times 10^{13} \ J \quad (3.5 \times 10^{13} \ J)}$$

The answer is (B).

2. The expression represents the work done in a rotational system by a variable torque.

$$W = \int \mathbf{T} \cdot d\theta$$

The answer is (D).

3. Because no value is given for the coil efficiency, assume 100%. Therefore, the work of the coil, W_{coil}, equals the work of a spring, W_{spring}, with spring constant, k, and positions, $\delta_{initial}$ and δ_{final}.

$$W_{coil} = W_{spring} = \tfrac{1}{2}k(\delta_{final}^2 - \delta_{initial}^2)$$

With no initial deflection, the equation becomes

$$W_{spring} = \tfrac{1}{2}k\delta_{final}^2$$
$$= \left(\tfrac{1}{2}\right)\left(15 \ \frac{N}{m}\right)(10 \ cm)^2 \left(\frac{1 \ m}{100 \ cm}\right)^2$$
$$= \boxed{0.075 \ N \cdot m}$$

The answer is (B).

4. The kinetic energy of the rotor is

$$E = \tfrac{1}{2}mv^2 \qquad [I]$$

The mass is m and the velocity is v.

The velocity is

$$v = \omega r$$
$$= \left(\frac{\left(1800 \ \frac{rev}{min}\right)\left(\frac{2\pi}{rev}\right)}{\left(60 \ \frac{s}{min}\right)}\right)(0.3 \ m)$$
$$= 56.55 \ m/s \qquad [II]$$

Substitute the result of Eq. II into Eq. I, along with the given information about the mass.

$$E = \tfrac{1}{2}mv^2$$
$$= \left(\tfrac{1}{2}\right)(100 \ kg)\left(56.55 \ \frac{m}{s}\right)^2$$
$$= \boxed{1.598 \times 10^5 \ J \quad (1.6 \times 10^5 \ J)}$$

This problem may also be solved using moment of inertia concepts.

The answer is (B).

5. During worst-case conditions, a motor's rotational energy will be converted to electrical energy and provided to the fault site. Power, in watts, is energy per unit time, or

$$P = \frac{E}{t}$$
$$= \frac{20 \times 10^4 \ J}{1 \ s}$$
$$= \boxed{20 \times 10^4 \ W \quad (200 \ kW)}$$

The answer is (C).

3 Engineering Drawing Practice

PRACTICE PROBLEMS

1. Which of the following symbols represents an electrical connection, or tie-point, between one part of an electrical circuit and another?

(A)

(B)

(C)

(D)

2. Which symbol represents a ground for safety purposes on a metallic enclosure?

(A)

(B)

(C)

(D)

3. Which of the following symbols would be used in an electrical schematic to represent a MOSFET?

(A)

(B)

(C)

(D)

4. Which of the following symbols would be used in an electrical schematic to represent an NPN bipolar junction transistor?

(A)

(B)

(C)

(D)

5. Which of the following symbols represents a controlled source?

(A)

(B)

(C)

(D)

Mathematics

SOLUTIONS

1. A tie-point is represented in an electrical schematic by the symbol

The answer is (A).

2. An earth ground is used for safety. The symbol for an earth ground is

$$\longrightarrow\!\!|\text{ı}\text{ı}$$

The answer is (B).

3. An *n*-channel enhancement (four-terminal) MOS-FET is symbolized by

The answer is (D).

4. An NPN bipolar junction transistor is represented by

A useful memory aid is that an NPN transistor's arrow is "<u>n</u>ot pointing <u>i</u>n" toward the base, giving NPN.

The answer is (A).

5. A controlled source is used in schematics of electronic models, and is represented by the diamond shape.

The answer is (C).

4 Algebra

PRACTICE PROBLEMS

Series

1. Calculate the following sum.

$$\sum_{j=1}^{5}\left((j+1)^2 - 1\right)$$

(A) 15

(B) 24

(C) 35

(D) 85

Logarithms

2. If every 0.1 sec a quantity increases by 0.1% of its current value, calculate the doubling time.

(A) 14 sec

(B) 70 sec

(C) 690 sec

(D) 69,000 sec

3. What are the roots of the following equation?

$$x^2 + 5x + 4 = 0$$

(A) $-1, -4$

(B) $-1, 4$

(C) $1, -4$

(D) $-11, -14$

4. What is the base-10 logarithm of 115^2?

(A) 2.06

(B) 4.12

(C) 8.24

(D) 16.5

5. What is the exponential form of $\mathbf{Z} = 2 + j2$?

(A) $0.7e^{j2.83}$

(B) $e^{j0.78}$

(C) $0.78e^{j}$

(D) $2.83e^{j0.78}$

SOLUTIONS

1. Let $S_n = (j+1)^2 - 1$.

For $j = 1$, $S_1 = (1+1)^2 - 1 = 3$

For $j = 2$, $S_2 = (2+1)^2 - 1 = 8$

For $j = 3$, $S_3 = (3+1)^2 - 1 = 15$

For $j = 4$, $S_4 = (4+1)^2 - 1 = 24$

For $j = 5$, $S_5 = (5+1)^2 - 1 = 35$

Substituting the above expressions gives

$$\sum_{j=1}^{5}\left((j+1)^2 - 1\right) = \sum_{j=1}^{5} S_j$$
$$= S_1 + S_2 + S_3 + S_4 + S_5$$
$$= 3 + 8 + 15 + 24 + 35$$
$$= \boxed{85}$$

The answer is (D).

2. Let n represent the number of elapsed periods of 0.1 sec, and let y_n represent the quantity present after n periods.

y_0 represents the initial quantity.

$$y_1 = 1.001 y_0$$
$$y_2 = 1.001 y_1 = (1.001)(1.001 y_0) = (1.001)^2 y_0$$

Therefore, by deduction,

$$y_n = (1.001)^n y_0$$

The expression for a doubling of the original quantity is

$$2 y_0 = y_n$$

Substitute for y_n.

$$2 y_0 = (1.001)^n y_0$$
$$2 = (1.001)^n$$

Take the logarithm of both sides.

$$\log 2 = \log 1.001^n$$
$$= n \log 1.001$$

Solve for n.

$$n = \frac{\log 2}{\log 1.001} = 693.5$$

Because each period, p, is 0.1 sec, the time is

$$t = np$$
$$= n(0.1 \text{ sec})$$
$$= (693.5)(0.1 \text{ sec})$$
$$= \boxed{69.35 \text{ sec} \quad (70 \text{ sec})}$$

The answer is (B).

3. The equation is in the quadratic form,

$$ax^2 + bx + c = 0$$

Quadratic equations can represent a second order response by electrical circuits. The roots are

$$x_1, x_2 = \frac{-b \pm \sqrt{b^2 - 4ac}}{2a}$$
$$= \frac{-5 \pm \sqrt{5^2 - (4)(1)(4)}}{(2)(1)}$$
$$= \boxed{-1, -4}$$

The answer is (A).

4. The property of logarithms used in this case is

$$\log x^a = a \log x$$

Substitute the given information.

$$\log 115^2 = 2 \log 115$$
$$= (2)(2.06)$$
$$= \boxed{4.12}$$

The answer is (B).

5. The exponential form of a complex number **Z** is

$$z = re^{j\theta}$$

The radius in the complex plane is r. The square root of -1 is j, and θ is the angle of the radius in the complex plane. Calculate the unknown values using

$$r = \sqrt{x_{real}^2 + y_{imaginary}^2}$$
$$= \sqrt{2^2 + 2^2}$$
$$= 2.83$$
$$\theta = \arctan \frac{y}{x}$$
$$= \arctan \frac{2}{2}$$
$$= 0.78 \text{ rad}$$

Therefore,

$$z = re^{j\theta}$$
$$= \boxed{2.83e^{j0.78}}$$

The answer is (D).

5 Linear Algebra

PRACTICE PROBLEMS

1. What is the determinant of matrix $\mathbf{A}$?

$$\mathbf{A} = \begin{bmatrix} 8 & 2 & 0 & 0 \\ 2 & 8 & 2 & 0 \\ 0 & 2 & 8 & 2 \\ 0 & 0 & 2 & 4 \end{bmatrix}$$

(A) 459

(B) 832

(C) 1552

(D) 1776

2. Use Cramer's rule to solve for the values of x, y, and z that simultaneously satisfy the following equations.

$$x + y = -4$$
$$x + z - 1 = 0$$
$$2z - y + 3x = 4$$

(A) $(x, y, z) = (3, 2, 1)$

(B) $(x, y, z) = (-3, -1, 2)$

(C) $(x, y, z) = (3, -1, -3)$

(D) $(x, y, z) = (-1, -3, 2)$

3. What value of x satisfies all three given equations?

$$4x + 6y - 8z = 2$$
$$6x - 2y - 4z = 8$$
$$8x - 14y - 12z = -14$$

(A) 0.10

(B) 0.20

(C) 0.33

(D) 3.0

4. What is the product of matrices $\mathbf{A}$ and $\mathbf{B}$?

$$\mathbf{A} = \begin{bmatrix} 1 & 2 & 5 \\ 3 & 4 & 7 \end{bmatrix}$$

$$\mathbf{B} = \begin{bmatrix} 2 & 9 \\ 6 & 2 \\ 3 & 4 \end{bmatrix}$$

(A) $\begin{bmatrix} 5 & 25 \\ 3 & 47 \end{bmatrix}$

(B) $\begin{bmatrix} 29 & 51 \\ 33 & 63 \end{bmatrix}$

(C) $\begin{bmatrix} 29 & 33 \\ 51 & 63 \end{bmatrix}$

(D) cannot be multiplied

5. What is the determinant of matrix $\mathbf{A}$?

$$\mathbf{A} = \begin{bmatrix} 4 & 4 \\ 3 & 6 \end{bmatrix}$$

(A) -12

(B) 12

(C) 24

(D) 36

SOLUTIONS

1. Expand by cofactors of the first column, because there are two zeros in that column. (The first row could also have been used.)

$$D = 8 \begin{vmatrix} 8 & 2 & 0 \\ 2 & 8 & 2 \\ 0 & 2 & 4 \end{vmatrix} - 2 \begin{vmatrix} 2 & 0 & 0 \\ 2 & 8 & 2 \\ 0 & 2 & 4 \end{vmatrix} + 0 - 0$$

by first column:

$$\begin{vmatrix} 8 & 2 & 0 \\ 2 & 8 & 2 \\ 0 & 2 & 4 \end{vmatrix} = (8)\big((8)(4) - (2)(2)\big)$$
$$- (2)\big((2)(4) - (2)(0)\big)$$
$$= 208$$

by first column:

$$\begin{vmatrix} 2 & 0 & 0 \\ 2 & 8 & 2 \\ 0 & 2 & 4 \end{vmatrix} = (2)\big((8)(4) - (2)(2)\big) = 56$$

$$D = (8)(208) - (2)(56) = \boxed{1552}$$

The answer is (C).

2. Rearrange the equations.

$$\begin{aligned} x + y \quad\;\; &= -4 \\ x \quad\;\; + z &= 1 \\ 3x - y + 2z &= 4 \end{aligned}$$

Write the set of equations in matrix form: $\mathbf{AX} = \mathbf{B}$.

$$\begin{bmatrix} 1 & 1 & 0 \\ 1 & 0 & 1 \\ 3 & -1 & 2 \end{bmatrix} \begin{bmatrix} x \\ y \\ z \end{bmatrix} = \begin{bmatrix} -4 \\ 1 \\ 4 \end{bmatrix}$$

Find the determinant of the matrix $\mathbf{A}$.

$$|\mathbf{A}| = \begin{vmatrix} 1 & 1 & 0 \\ 1 & 0 & 1 \\ 3 & -1 & 2 \end{vmatrix}$$

$$= 1 \begin{vmatrix} 0 & 1 \\ -1 & 2 \end{vmatrix} - 1 \begin{vmatrix} 1 & 0 \\ -1 & 2 \end{vmatrix} + 3 \begin{vmatrix} 1 & 0 \\ 0 & 1 \end{vmatrix}$$

$$= (1)\big((0)(2) - (1)(-1)\big)$$
$$- (1)\big((1)(2) - (-1)(0)\big)$$
$$+ (3)\big((1)(1) - (0)(0)\big)$$
$$= (1)(1) - (1)(2) + (3)(1)$$
$$= 1 - 2 + 3$$
$$= 2$$

Find the determinant of the substitutional matrix $\mathbf{A}_1$.

$$|\mathbf{A}_1| = \begin{vmatrix} -4 & 1 & 0 \\ 1 & 0 & 1 \\ 4 & -1 & 2 \end{vmatrix}$$

$$= -4 \begin{vmatrix} 0 & 1 \\ -1 & 2 \end{vmatrix} - 1 \begin{vmatrix} 1 & 0 \\ -1 & 2 \end{vmatrix} + 4 \begin{vmatrix} 1 & 0 \\ 0 & 1 \end{vmatrix}$$

$$= (-4)\big((0)(2) - (1)(-1)\big)$$
$$- (1)\big((1)(2) - (-1)(0)\big)$$
$$+ (4)\big((1)(1) - (0)(0)\big)$$
$$= (-4)(1) - (1)(2) + (4)(1)$$
$$= -4 - 2 + 4$$
$$= -2$$

Find the determinant of the substitutional matrix $\mathbf{A}_2$.

$$|\mathbf{A}_2| = \begin{vmatrix} 1 & -4 & 0 \\ 1 & 1 & 1 \\ 3 & 4 & 2 \end{vmatrix}$$

$$= 1 \begin{vmatrix} 1 & 1 \\ 4 & 2 \end{vmatrix} - 1 \begin{vmatrix} -4 & 0 \\ 4 & 2 \end{vmatrix} + 3 \begin{vmatrix} -4 & 0 \\ 1 & 1 \end{vmatrix}$$

$$= (1)\big((1)(2) - (4)(1)\big)$$
$$- (1)\big((-4)(2) - (4)(0)\big)$$
$$+ (3)\big((-4)(1) - (1)(0)\big)$$
$$= (1)(-2) - (1)(-8) + (3)(-4)$$
$$= -2 + 8 - 12$$
$$= -6$$

Find the determinant of the substitutional matrix $\mathbf{A}_3$.

$$|\mathbf{A}_3| = \begin{vmatrix} 1 & 1 & -4 \\ 1 & 0 & 1 \\ 3 & -1 & 4 \end{vmatrix}$$

$$= 1\begin{vmatrix} 0 & 1 \\ -1 & 4 \end{vmatrix} - 1\begin{vmatrix} 1 & -4 \\ -1 & 4 \end{vmatrix} + 3\begin{vmatrix} 1 & -4 \\ 0 & 0 \end{vmatrix}$$

$$= (1)\big((0)(4) - (-1)(1)\big)$$
$$\quad - (1)\big((1)(4) - (-1)(-4)\big)$$
$$\quad + (3)\big((1)(1) - (0)(-4)\big)$$
$$= (1)(1) - (1)(0) + (3)(1)$$
$$= 1 - 0 + 3$$
$$= 4$$

Use Cramer's rule.

$$x = \frac{|\mathbf{A}_1|}{|\mathbf{A}|} = \frac{-2}{2} = \boxed{-1}$$
$$y = \frac{|\mathbf{A}_2|}{|\mathbf{A}|} = \frac{-6}{2} = \boxed{-3}$$
$$z = \frac{|\mathbf{A}_3|}{|\mathbf{A}|} = \frac{4}{2} = \boxed{2}$$

The answer is (D).

3. Use Cramer's rule to solve for x. The determinant of the coefficient matrix is

$$|\mathbf{A}| = \begin{vmatrix} 4 & 6 & -8 \\ 6 & -2 & -4 \\ 8 & -14 & -12 \end{vmatrix}$$

$$= 4\begin{vmatrix} -2 & -4 \\ -14 & -12 \end{vmatrix} - 6\begin{vmatrix} 6 & -8 \\ -14 & -12 \end{vmatrix} + 8\begin{vmatrix} 6 & -8 \\ -2 & -4 \end{vmatrix}$$

$$= (4)\big((-2)(-12) - (-4)(-14)\big)$$
$$\quad - (6)\big((6)(-12) - (-8)(-14)\big)$$
$$\quad + (8)\big((6)(-4) - (-2)(-8)\big)$$
$$= (4)(24 - 56) - (6)(-72 - 112)$$
$$\quad + (8)(-24 - 16)$$
$$= 656$$

The determinant of the substitutional matrix is

$$|\mathbf{A}_1| = \begin{vmatrix} 2 & 6 & -8 \\ 8 & -2 & -4 \\ -14 & -14 & -12 \end{vmatrix}$$
$$= 1968$$

Therefore,

$$x = \frac{|\mathbf{A}_1|}{|\mathbf{A}|} = \frac{1968}{656} = 3$$
$$= \boxed{3}$$

The answer is (D).

4. Let the product matrix be $\mathbf{C}$.

$$\mathbf{C} = \begin{bmatrix} c_{11} & c_{12} \\ c_{21} & c_{22} \end{bmatrix} = \mathbf{AB}$$

$$= \begin{bmatrix} 1 & 2 & 5 \\ 3 & 4 & 7 \end{bmatrix}\begin{bmatrix} 2 & 9 \\ 6 & 2 \\ 3 & 4 \end{bmatrix}$$

$$c_{11} = (1)(2) + (2)(6) + (5)(3)$$
$$= 2 + 12 + 15$$
$$= 29$$
$$c_{12} = (1)(9) + (2)(2) + (5)(4)$$
$$= 9 + 4 + 20$$
$$= 33$$
$$c_{21} = (3)(2) + (4)(6) + (7)(3)$$
$$= 6 + 24 + 21$$
$$= 51$$
$$c_{22} = (3)(9) + (4)(2) + (7)(4)$$
$$= 27 + 8 + 28$$
$$= 63$$

Therefore,

$$\mathbf{C} = \boxed{\begin{bmatrix} 29 & 33 \\ 51 & 63 \end{bmatrix}}$$

The answer is (C).

5. The determinant of any 2×2 matrix is

$$|\mathbf{A}| = \begin{vmatrix} a & c \\ b & d \end{vmatrix} = ad - bc$$

In this case,

$$\begin{aligned} |\mathbf{A}| &= \begin{vmatrix} 4 & 4 \\ 3 & 6 \end{vmatrix} = (4)(6) - (4)(3) \\ &= 24 - 12 \\ &= \boxed{12} \end{aligned}$$

The answer is (B).

6 Vectors

PRACTICE PROBLEMS

1. What is the unit vector for $\mathbf{V} = 2\mathbf{i} + 4\mathbf{j} + 4\mathbf{k}$?

(A) $1\mathbf{i} + 2\mathbf{j} + 2\mathbf{k}$

(B) $(1/18)\mathbf{i} + (1/6)\mathbf{j} + (1/6)\mathbf{k}$

(C) $(2/3)\mathbf{i} + (1/3)\mathbf{j} + (1/3)\mathbf{k}$

(D) $(1/3)\mathbf{i} + (2/3)\mathbf{j} + (2/3)\mathbf{k}$

2. What is the dot product of the two vectors given?

$$\mathbf{V}_1 = 2\mathbf{i} + 2\mathbf{j} + 4\mathbf{k}$$
$$\mathbf{V}_2 = 2\mathbf{i} + 2\mathbf{j} + 1\mathbf{k}$$

(A) $4\angle 35°$

(B) 4

(C) $12\angle 35°$

(D) 12

3. Which of the following vectors is orthogonal to $\mathbf{V}_1$ and $\mathbf{V}_2$?

$$\mathbf{V}_1 = \mathbf{i} + \mathbf{j} + 2\mathbf{k}$$
$$\mathbf{V}_2 = \mathbf{i} + 2\mathbf{j}$$

(A) $-\mathbf{i} - 2\mathbf{j} + 2\mathbf{k}$

(B) $-4\mathbf{i} + 2\mathbf{j} + \mathbf{k}$

(C) $-4\mathbf{i} - 2\mathbf{j} - \mathbf{k}$

(D) $4\mathbf{i} - 2\mathbf{j} + \mathbf{k}$

4. The magnitude of the vector cross product corresponds to which of the following?

(A) area

(B) line

(C) projection

(D) volume

5. Which of the following vectors is orthogonal to $\mathbf{V}_1 = 2\mathbf{i} + 4\mathbf{j}$?

(A) $2\mathbf{i} - 4\mathbf{j}$

(B) $4\mathbf{i} - 2\mathbf{j}$

(C) $-2\mathbf{i} - 4\mathbf{j}$

(D) $2\mathbf{i} + 4\mathbf{j}$

SOLUTIONS

1. The unit vector $\mathbf{a}$ for vector $\mathbf{V}$ is

$$
\begin{aligned}
\mathbf{a} &= \frac{\mathbf{V}}{|\mathbf{V}|} \\
&= \frac{2\mathbf{i} + 4\mathbf{j} + 4\mathbf{k}}{\sqrt{2^2 + 4^2 + 4^2}} \\
&= \frac{2\mathbf{i} + 4\mathbf{j} + 4\mathbf{k}}{6} \\
&= \boxed{(1/3)\mathbf{i} + (2/3)\mathbf{j} + (2/3)\mathbf{k}}
\end{aligned}
$$

The answer is (D).

2. The dot product is

$$
\begin{aligned}
\mathbf{V}_1 \cdot \mathbf{V}_2 &= |\mathbf{V}_1||\mathbf{V}_2|\cos\phi = V_{1x}V_{2x} + V_{1y}V_{2y} + V_{1z}V_{2z} \\
&= (2)(2) + (2)(2) + (4)(1) \\
&= \boxed{12}
\end{aligned}
$$

The answer is (D).

3. An orthogonal vector is found from the cross product. The order of the cross product will change the perpendicular found. Using $\mathbf{V}_1 \times \mathbf{V}_2$ gives

$$
\begin{aligned}
\mathbf{V}_1 \times \mathbf{V}_2 &= \begin{vmatrix} \mathbf{i} & 1 & 1 \\ \mathbf{j} & 1 & 2 \\ \mathbf{k} & 2 & 0 \end{vmatrix} \\
&= \mathbf{i}\begin{vmatrix} 1 & 2 \\ 2 & 0 \end{vmatrix} - \mathbf{j}\begin{vmatrix} 1 & 1 \\ 2 & 0 \end{vmatrix} + \mathbf{k}\begin{vmatrix} 1 & 1 \\ 1 & 2 \end{vmatrix} \\
&= \mathbf{i}(0 - 4) - \mathbf{j}(0 - 2) + \mathbf{k}(2 - 1) \\
&= \boxed{-4\mathbf{i} + 2\mathbf{j} + \mathbf{k}}
\end{aligned}
$$

The answer is (B).

4. The magnitude of the vector cross product for two vectors, say $\mathbf{V}_1 \times \mathbf{V}_2$, corresponds to the area of a parallelogram with $\mathbf{V}_1$ and $\mathbf{V}_2$ as its sides.

The answer is (A).

5. A vector is orthogonal to another if the dot product of the two vectors equals zero.

Checking answer (A) gives

$$
\begin{aligned}
\mathbf{V}_1 \cdot \mathbf{A} &= V_{1x}A_{1x} + V_{2y}A_{2y} \\
&= (2)(2) + (4)(-4) \\
&= -12
\end{aligned}
$$

Because the dot product result is not zero, $\mathbf{A}$ is not orthogonal.

Checking answer (B) gives

$$
\begin{aligned}
\mathbf{V}_1 \cdot \mathbf{B} &= V_{1x}B_{1x} + V_{2y}B_{2y} \\
&= (2)(4) + (4)(-2) \\
&= 0
\end{aligned}
$$

$$\boxed{4\mathbf{i} - 2\mathbf{j} \text{ is orthogonal to } \mathbf{V}_1 = 2\mathbf{i} + 4\mathbf{j}.}$$

The answer is (B).

7 Trigonometry

PRACTICE PROBLEMS

1. A 5 lbm (5 kg) block sits on a 20° incline without slipping.

(a) Draw the free-body diagram with respect to axes parallel and perpendicular to the surface of the incline.

(b) Determine the magnitude of the frictional force on the block.

    (A) 1.71 lbf (16.8 N)

    (B) 3.35 lbf (32.9 N)

    (C) 4.70 lbf (46.1 N)

    (D) 5.00 lbf (49.1 N)

2. Consider the unit circle shown.

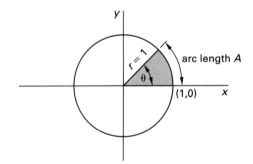

What is the value of the arc length, A, in radians, if the shaded area is $\theta/2$?

    (A) θ

    (B) $\pi/4$

    (C) $\pi/2$

    (D) $r\theta$

3. If the cosine of the angle θ is given as C, what is the tangent of the angle θ?

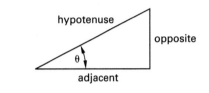

    (A) $\dfrac{C \times \text{hypotenuse}}{\text{opposite}}$

    (B) $\dfrac{\text{adjacent}}{\text{opposite}}$

    (C) $\dfrac{\text{opposite} \times \text{adjacent}}{C \times \text{hypotenuse}}$

    (D) $\dfrac{\text{opposite}}{C \times \text{hypotenuse}}$

4. What is the value of the sum of the squares of the sine and cosine of an angle θ?

    (A) 0

    (B) $\pi/4$

    (C) 1

    (D) 2π

5. What is the approximate value of tan 10° in radians?

    (A) 1.00

    (B) 10.0°

    (C) 0.0500

    (D) 0.175

SOLUTIONS

1. (a)

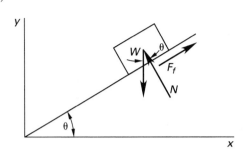

Customary U.S. Solution

(b) The mass of the block, m, is 5 lbm.

The angle of inclination, θ, is 20°. The weight is

$$W = \frac{mg}{g_c}$$

$$= \frac{(5 \text{ lbm})\left(32.2 \ \frac{\text{ft}}{\text{sec}^2}\right)}{32.2 \ \frac{\text{lbm-ft}}{\text{lbf-sec}^2}}$$

$$= 5 \text{ lbf}$$

The frictional force is

$$F_f = W \sin \theta$$

$$= (5 \text{ lbf}) \sin 20°$$

$$= \boxed{1.71 \text{ lbf}}$$

The answer is (A).

SI Solution

(b) The mass of the block, m, is 5 kg.

The angle of inclination, θ, is 20°. The gravitational force is

$$W = mg$$

$$= (5 \text{ kg})\left(9.81 \ \frac{\text{m}}{\text{s}^2}\right)$$

$$= 49.1 \text{ N}$$

The frictional force is

$$F_f = W \sin \theta$$

$$= (49.1 \text{ N}) \sin 20°$$

$$= \boxed{16.8 \text{ N}}$$

The answer is (A).

2. The area of a sector with a central angle of θ radians is $\theta/2$ units for a unit circle. The arc length is therefore θ radians.

The answer is (A).

3. The cosine is

$$C = \frac{\text{adjacent}}{\text{hypotenuse}} \qquad \text{[I]}$$

The tangent is

$$T = \frac{\text{opposite}}{\text{adjacent}} \qquad \text{[II]}$$

Substituting the value of adjacent from Eq. I into Eq. II gives

$$T = \frac{\text{opposite}}{\text{adjacent}} = \boxed{\frac{\text{opposite}}{C \times \text{hypotenuse}}}$$

The answer is (D).

4. For any angle θ, the following is true.

$$\cos^2 \theta + \sin^2 \theta = \boxed{1}$$

The answer is (C).

5. For angles less than approximately 10°, the following is valid.

$$\sin \theta \approx \tan \theta \approx \theta\big|_{\theta \leq 10°}$$

The value of θ must be expressed in radians. Converting the 10° to radians gives

$$\theta_{\text{radians}} = 10°\left(\frac{2\pi \text{ rad}}{360°}\right)$$

$$= 0.05\pi \text{ rad}$$

$$= \boxed{0.175 \text{ rad}}$$

The answer is (D).

8 Analytic Geometry

PRACTICE PROBLEMS

1. The diameter of a sphere and of the base of a cone are equal. What percentage of that diameter must the cone's height be, so that both volumes are equal?

(A) 133%

(B) 150%

(C) 166%

(D) 200%

2. Consider the graph shown. What is the slope of the line?

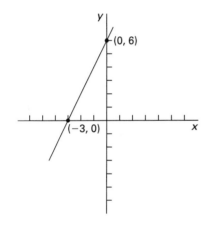

(A) −1/2

(B) 1/2

(C) 2

(D) 4

3. What type of symmetry is shown?

(A) even

(B) odd

(C) rotational

(D) quadratic

4. What expression in the spherical coordinate system represents the z coordinate in the Cartesian coordinate system?

(A) z

(B) $r \sin \theta$

(C) $r \cos \phi$

(D) $r \sin \phi \sin \theta$

5. What geometric figure does the following equation represent?

$$(x - 2)^2 + (y - 2)^2 + (z - 2)^2 = 100$$

(A) circle

(B) ellipse

(C) helix

(D) sphere

SOLUTIONS

1. Let d be the diameter of the sphere and the base of the cone.

The volume of the sphere is

$$V_{\text{sphere}} = \frac{4}{3}\pi r^3 = \frac{4}{3}\pi\left(\frac{d}{2}\right)^3$$
$$= \frac{\pi}{6}d^3$$

The volume of the circular cone is

$$V_{\text{cone}} = \frac{1}{3}\pi r^2 h = \frac{1}{3}\pi\left(\frac{d}{2}\right)^2 h$$
$$= \frac{\pi}{12}d^2 h$$

The volumes of the sphere and cone are equal.

$$V_{\text{cone}} = V_{\text{sphere}}$$
$$\frac{\pi}{12}d^2 h = \frac{\pi}{6}d^3$$
$$h = 2d$$

The cone's height must be 200% of its diameter.

The answer is (D).

2. The two-point form of a line is

$$\frac{y - y_1}{x - x_1} = \frac{y_2 - y_1}{x_2 - x_1}$$

Substitute for the two points given.

$$\frac{y - 6}{x - 0} = \frac{0 - 6}{-3 - 0}$$
$$\frac{y - 6}{x} = 2$$
$$y - 6 = 2x$$
$$y = 2x + 6$$

The equation is in the slope intercept form of a line,

$$y = mx + b$$

The slope value is m. Therefore, in this case, the slope is 2.

The answer is (C).

3. The symmetry is such that the value of y is identical for any $\pm x$ value. This is even symmetry.

The answer is (A).

4. In the spherical system

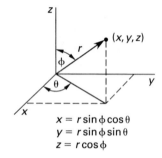

$x = r\sin\phi\cos\theta$
$y = r\sin\phi\sin\theta$
$z = r\cos\phi$

The value of z is $r\cos\phi$.

The answer is (C).

5. The general equation for a sphere that is centered at (h, k, l) is

$$(x - h)^2 + (y - k)^2 + (z - l)^2 = r^2$$

The equation given is in this form. The equation represents a sphere centered at $(2, 2, 2)$.

The answer is (D).

9 Differential Calculus

PRACTICE PROBLEMS

1. What are the values of a, b, and c in the following expression, such that $n(\infty) = 100$, $n(0) = 10$, and $dn(0)/dt = 0.5$?

$$n(t) = \frac{a}{1 + be^{ct}}$$

(A) $a = 10$, $b = 9$, $c = 1$

(B) $a = 100$, $b = 10$, $c = 1.5$

(C) $a = 100$, $b = 9$, $c = -0.056$

(D) $a = 1000$, $b = 10$, $c = -0.056$

2. Find all minima, maxima, and inflection points for

$$y = x^3 - 9x^2 - 3$$

(A)   maximum at $x = 0$
       inflection at $x = 3$
       minimum at $x = 6$

(B)   maximum at $x = 0$
       inflection at $x = -3$
       minimum at $x = -6$

(C)   maximum at $x = 6$
       inflection at $x = 0$
       minimum at $x = -3$

(D)   maximum at $x = -3$
       inflection at $x = 0$
       minimum at $x = 3$

3. What is the derivative of the following equation?

$$f(t) = 170 \cos 377t$$

(A) $120 \sin 377t$

(B) $170 \sin 377t$

(C) $-170 \sin 377t$

(D) $-64{,}000 \sin 377t$

4. What is the gradient of the following function?

$$f(x, y) = x^2 - 2y^2 + 2x + y$$

(A) $(2x - 2)\mathbf{i} - (4y + 1)\mathbf{j}$

(B) $(2x + 2)\mathbf{i} - (-4y + 1)\mathbf{j}$

(C) $(-2x - 2)\mathbf{i} + (-4y + 1)\mathbf{j}$

(D) $(2x + 2)\mathbf{i} + (-4y + 1)\mathbf{j}$

5. What is the divergence of the following vector function?

$$\mathbf{F}(x, y, z) = 2xz\mathbf{i} + e^x y^2\mathbf{j} + x^2 y\mathbf{k}$$

(A) $(2z + e^x + 2x)\mathbf{i} + (2ye^x + x^2)\mathbf{j} + 2x\mathbf{k}$

(B) $2z + 2ye^x$

(C) $2z + 2ye^x + 1$

(D) $2z\mathbf{i} + 2ye^x\mathbf{j} + \mathbf{k}$

SOLUTIONS

1. If c is positive, then $n(\infty)$ is zero, which is contrary to the data given. Therefore, c is less than or equal to zero. If c is zero, then $n(\infty) = a/(1+b) = 100$, which is possible depending on the values of a and b. However, $n(0) = a/(1+b)$ would also equal 100, which is contrary to the given data. Therefore, c is not zero. c must be less than zero.

Because $c < 0$, then $n(\infty) = a$, so $\boxed{a = 100.}$

Applying the condition $t = 0$ gives

$$n(0) = \frac{a}{1+b} = 10$$

Because $a = 100$,

$$n(0) = \frac{100}{1+b}$$
$$100 = (10)(1+b)$$
$$10 = 1 + b$$
$$\boxed{b = 9}$$

Substitute the results for a and b into the expression.

$$n(t) = \frac{100}{1 + 9e^{ct}}$$

Take the first derivative.

$$\frac{d}{dt}n(t) = \left(\frac{100}{(1+9e^{ct})^2}\right)(-9ce^{ct})$$

Apply the initial condition.

$$\frac{d}{dt}n(0) = \left(\frac{100}{(1+9e^{c(0)})^2}\right)(-9ce^{c(0)}) = 0.5$$

$$\left(\frac{100}{(1+9)^2}\right)(-9c) = 0.5$$

$$(1)(-9c) = 0.5$$

$$c = \frac{-0.5}{9}$$

$$= \boxed{-0.056}$$

Substitute the terms a, b, and c into the expression.

$$n(t) = \frac{100}{1 + 9e^{-0.056t}}$$

The answer is (C).

2. Determine the critical points by taking the first derivative of the function and setting it equal to zero.

$$\frac{dy}{dx} = 3x^2 - 18x = 3x(x-6)$$

$$3x(x-6) = 0$$

$$x(x-6) = 0$$

The critical points are located at $x = 0$ and $x = 6$.

Determine the inflection points by setting the second derivative equal to zero. Take the second derivative.

$$\frac{d^2y}{dx^2} = \left(\frac{d}{dx}\right)\left(\frac{dy}{dx}\right) = \frac{d}{dx}(3x^2 - 18x)$$

$$= 6x - 18$$

Set the second derivative equal to zero.

$$\frac{d^2y}{dx^2} = 0 = 6x - 18 = 6(x-3)$$

$$6(x-3) = 0$$

$$x - 3 = 0$$

$$x = 3$$

$\boxed{\text{This inflection point is at } x = 3.}$

Determine the local maximum and minimum by substituting the critical points into the expression for the second derivative.

At the critical point $x = 0$,

$$\left.\frac{d^2y}{dx^2}\right|_{x=0} = 6(x-3) = (6)(0-3)$$

$$= -18$$

Because $-18 < 0$, $\boxed{x = 0 \text{ is a local maximum.}}$

At the critical point $x = 6$,

$$\left.\frac{d^2y}{dx^2}\right|_{x=6} = 6(x-3) = (6)(6-3)$$

$$= 18$$

Because $18 > 0$, $\boxed{x = 6 \text{ is a local minimum.}}$

The answer is (A).

3. The derivative of $f(t)$ is

$$f'(t) = 170 \frac{d \cos 377t}{dt}$$

$$= (170)(-\sin 377t)\left(\frac{d(377t)}{dt}\right)$$

$$= (-170)\sin 377t \left(377\frac{dt}{dt}\right)$$

$$= \boxed{-64{,}090 \sin 377t \quad (-64{,}000 \sin 377t)}$$

The answer is (D).

4. The gradient of $f(x, y)$ is

$$\nabla f(x, y) = \frac{\partial f(x, y)}{\partial x}\mathbf{i} + \frac{\partial f(x, y)}{\partial y}\mathbf{j}$$

$$= \frac{\partial(x^2 - 2y^2 + 2x + y)}{\partial x}\mathbf{i}$$

$$+ \frac{\partial(x^2 - 2y^2 + 2x + y)}{\partial y}\mathbf{j}$$

$$= \boxed{(2x + 2)\mathbf{i} + (-4y + 1)\mathbf{j}}$$

The answer is (D).

5. The divergence is a scalar, given by

$$\text{div } \mathbf{F} = \nabla \cdot \mathbf{F} = \frac{\partial F_x}{\partial x} + \frac{\partial F_y}{\partial y} + \frac{\partial F_z}{\partial z}$$

$$= \frac{\partial(2xz)}{\partial x} + \frac{\partial(e^x y^2)}{\partial y} + \frac{\partial(x^2 y)}{\partial z}$$

$$= 2z + 2e^x y + 0$$

$$= \boxed{2z + 2ye^x}$$

The answer is (B).

10 Integral Calculus

PRACTICE PROBLEMS

1. What is the result of the indefinite integration shown?

$$\int \sin \theta \, d\theta$$

(A) $-\cos \theta + C$

(B) $\cos \theta + C$

(C) $-\sin \theta + C$

(D) $\dfrac{-\sin^2 \theta}{2}$

2. What is the value of the following definite integral?

$$\int_0^{\pi/2} \sin x \, dx$$

(A) -1

(B) 0

(C) 1

(D) $\pi/2$

3. Consider the following square wave function.

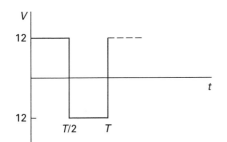

The period of the function is 0.0167 s. What is the value of a_0? That is, what is the value of the first term in a Fourier series representing the function?

(A) -6

(B) 0

(C) $+6$

(D) $+12$

4. What is the area between the curve shown, $y_1 = e^x$, and the straight line, $y_2 = x + 1$, shown on the interval $x = [0, 2]$?

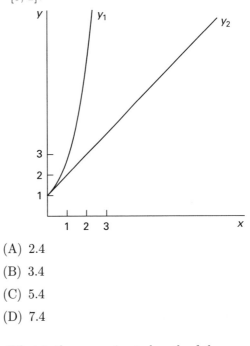

(A) 2.4

(B) 3.4

(C) 5.4

(D) 7.4

5. What is the approximate length of the curve $y = x^2$ from $x = 0$ to $x = 5$?

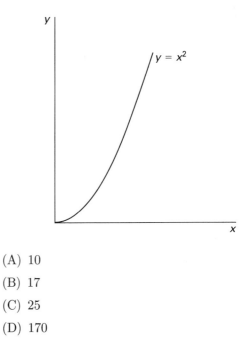

(A) 10

(B) 17

(C) 25

(D) 170

Mathematics

SOLUTIONS

1. The indefinite integral of the sine function is

$$\int \sin\theta\, d\theta = \boxed{-\cos\theta + C}$$

C represents a constant.

The answer is (A).

2. The definite integral is

$$\int_0^{\pi/2} \sin x\, dx = -\cos x \Big|_0^{\pi/2}$$

$$= -\cos\frac{\pi}{2} - (-\cos 0)$$

$$= 0 + 1$$

$$= \boxed{1}$$

The answer is (C).

3. The Fourier series is

$$f(t) = \tfrac{1}{2}a_0 + a_1\cos\omega t + a_2\cos 2\omega t + \cdots$$
$$+ b_1\sin\omega t + b_2\sin 2\omega t + \cdots$$

Normalizing by dividing all frequencies by ω gives

$$f(t) = \tfrac{1}{2}a_0 + a_1\cos t + a_2\cos 2t + \cdots$$
$$+ b_1\sin t + b_2\sin 2t + \cdots$$

The first term, $(1/2)a_0$, is the average value of the function. Calculate the value of a_0.

$$a_0 = \frac{2}{T}\int_0^T f(t)\, dt$$

$$= \frac{2}{T}\left(\int_0^{T/2} 12\, dt + \int_{T/2}^T (-12)\, dt\right)$$

$$= \frac{2}{T}\left(12t\big|_0^{T/2} - 12t\big|_{T/2}^T\right)$$

$$= \frac{2}{T}\left(\left(12\frac{T}{2} - (12)(0)\right) + \left((-12T) - (-12)\frac{T}{2}\right)\right)$$

$$= \frac{2}{T}\left(12\frac{T}{2} - 12T + 12\frac{T}{2}\right)$$

$$= \frac{2}{T}(12T - 12T)$$

$$= \frac{2}{T}(0)$$

$$= \boxed{0}$$

The average value, $(1/2)a_0$, equals zero, which can also be determined by inspection of the waveform.

The answer is (B).

4. The area bounded by the two functions is given by the following.

$$A = \int_a^b \big(y_1(x) - y_2(x)\big)\, dx$$

$$= \int_0^2 \big(e^x - (x+1)\big)\, dx = \int_0^2 (e^x - x - 1)\, dx$$

$$= \int_0^2 e^x\, dx - \int_0^2 x\, dx - \int_0^2 dx$$

$$= e^x \Big|_0^2 - \frac{x^2}{2}\Big|_0^2 - x\Big|_0^2$$

$$= (e^2 - e^0) - \left(\frac{2^2}{2} - \frac{0^2}{2}\right) - (2 - 0)$$

$$= e^2 - 1 - 2 - 2 = e^2 - 5 = 7.389 - 5$$

$$= \boxed{2.389 \quad (2.4)}$$

The answer is (A).

5. The length of the curve may be found from

$$L_{\text{curve}} = \int_a^b \sqrt{1 + \big(f'(x)\big)^2}\, dx \qquad \text{[I]}$$

Because $f(x) = x^2$, the derivative $f'(x) = 2x$. Substitute the derivative function and the given limits into Eq. I.

$$L_{\text{curve}} = \int_0^5 \sqrt{1 + (2x)^2}\, dx$$

$$= \int_0^5 \sqrt{1 + 4x^2}\, dx \qquad \text{[II]}$$

Use a table of integrals to determine the following.

$$\int \sqrt{a^2 + x^2}\, dx = \tfrac{1}{2}\left(\left(x\sqrt{a^2 + x^2}\right)\right.$$
$$\left.+ a^2\ln\left(x + \sqrt{a^2 + x^2}\right)\right)$$

Put Eq. II into the form of the solution from the table of integrals and solve.

$$L_{\text{curve}} = \int_0^5 \sqrt{1 + 4x^2}\,dx = 2\int_0^5 \sqrt{\left(\tfrac{1}{2}\right)^2 + x^2}\,dx$$

$$= (2)\left[\left(\tfrac{1}{2}\right)\left(\begin{array}{c} x\sqrt{\left(\tfrac{1}{2}\right)^2 + x^2} + \left(\tfrac{1}{2}\right)^2 \\ \times \ln\left(x + \sqrt{\left(\tfrac{1}{2}\right)^2 + x^2}\right) \end{array}\right)\Bigg|_0^5\right]$$

$$= (2)\left(\begin{array}{c} \left(\tfrac{1}{2}\right)\left(\begin{array}{c} 5\sqrt{\left(\tfrac{1}{2}\right)^2 + 5^2} + \left(\tfrac{1}{2}\right)^2 \\ \times \ln\left(5 + \sqrt{\left(\tfrac{1}{2}\right)^2 + 5^2}\right) \end{array}\right) \\ - \left(\tfrac{1}{2}\right)\left(\begin{array}{c} 0\sqrt{\left(\tfrac{1}{2}\right)^2 + 0^2} + \left(\tfrac{1}{2}\right)^2 \\ \times \ln\left(0 + \sqrt{\left(\tfrac{1}{2}\right)^2 + 0^2}\right) \end{array}\right) \end{array}\right)$$

$$= \boxed{25.8 \quad (25)}$$

The answer is (C).

11 Differential Equations

PRACTICE PROBLEMS

1. Solve the following differential equation for y.

$$y'' - 4y' - 12y = 0$$

(A) $A_1 e^{6x} + A_2 e^{-2x}$

(B) $A_1 e^{-6x} + A_2 e^{2x}$

(C) $A_1 e^{6x} + A_2 e^{2x}$

(D) $A_1 e^{-6x} + A_2 e^{-2x}$

2. Solve the following differential equation for y.

$$y' - y = 2xe^{2x} \qquad y(0) = 1$$

(A) $y = 2e^{-2x}(x-1) + 3e^{-x}$

(B) $y = 2e^{2x}(x-1) + 3e^{x}$

(C) $y = -2e^{-2x}(x-1) + 3e^{-x}$

(D) $y = 2e^{2x}(x-1) + 3e^{-x}$

3. The oscillation exhibited by the top story of a certain building in free motion is given by the following differential equation.

$$x'' + 2x' + 2x = 0 \qquad x(0) = 0 \qquad x'(0) = 1$$

(a) What is x as a function of time?

(A) $e^{-2t} \sin t$

(B) $e^{t} \sin t$

(C) $e^{-t} \sin t$

(D) $e^{-t} \sin t + e^{-t} \cos t$

(b) What is the building's fundamental natural frequency of vibration?

(A) $^1/_2$

(B) 1

(C) $\sqrt{2}$

(D) 2

(c) What is the amplitude of oscillation?

(A) 0.32

(B) 0.54

(C) 1.7

(D) 6.6

(d) What is x as a function of time if a lateral wind load is applied with a form of $\sin t$?

(A) $\frac{2}{5}e^{-t}\cos t + \frac{6}{5}e^{-t}\sin t$

(B) $\frac{2}{5}e^{t}\cos t + \frac{6}{5}e^{t}\sin t$

(C) $\frac{6}{5}e^{-t}\cos t + \frac{2}{5}e^{-t}\sin t - \frac{2}{5}\cos t + \frac{2}{5}\sin t$

(D) $\frac{2}{5}e^{-t}\cos t + \frac{6}{5}e^{-t}\sin t - \frac{2}{5}\cos t + \frac{1}{5}\sin t$

4. (*Time limit: one hour*) A 90 lbm (40 kg) bag of a chemical is accidentally dropped into an aerating lagoon. The chemical is water soluble and nonreacting. The lagoon is 120 ft (35 m) in diameter and filled to a depth of 10 ft (3 m). The aerators circulate and distribute the chemical evenly throughout the lagoon.

Water enters the lagoon at a rate of 30 gal/min (115 L/min). Fully mixed water is pumped into a reservoir at a rate of 30 gal/min (115 L/min).

The established safe concentration of this chemical is 1 ppb (part per billion). Approximately how many days will it take for the concentration of the discharge water to reach this level?

(A) 25 days

(B) 50 days

(C) 100 days

(D) 200 days

5. A tank contains 100 gal (100 L) of brine made by dissolving 60 lbm (60 kg) of salt in pure water. Salt water with a concentration of 1 lbm/gal (1 kg/L) enters the tank at a rate of 2 gal/min (2 L/min). A well-stirred mixture is drawn from the tank at a rate of 3 gal/min (3 L/min). Find the mass of salt in the tank after 1 hr.

(A) 13 lbm (13 kg)

(B) 37 lbm (37 kg)

(C) 43 lbm (43 kg)

(D) 51 lbm (51 kg)

SOLUTIONS

1. Obtain the characteristic equation by replacing each derivative with a polynomial term of equal degree.

$$r^2 - 4r - 12 = 0$$

Factor the characteristic equation.

$$(r - 6)(r + 2) = 0$$

The roots are $r_1 = 6$ and $r_2 = -2$.

Since the roots are real and distinct, the solution is

$$y = A_1 e^{r_1 x} + A_2 e^{r_2 x}$$
$$= \boxed{A_1 e^{6x} + A_2 e^{-2x}}$$

The answer is (A).

2. The equation is a first-order linear differential equation of the form

$$y' + p(x)y = g(x)$$
$$p(x) = -1$$
$$g(x) = 2xe^{2x}$$

The integration factor $u(x)$ is given by

$$u(x) = e^{\left(\int p(x)\,dx\right)}$$
$$= e^{\left(\int (-1)\,dx\right)}$$
$$= e^{-x}$$

The closed form of the solution is given by

$$y = \left(\frac{1}{u(x)}\right)\left(\int u(x)g(x)\,dx + C\right)$$
$$= \left(\frac{1}{e^{-x}}\right)\left(\int (e^{-x})(2xe^{2x})\,dx + C\right)$$
$$= e^x\left(2(xe^x - e^x) + C\right)$$
$$= e^x\left(2e^x(x - 1) + C\right)$$
$$= 2e^{2x}(x - 1) + Ce^x$$

Apply the initial condition $y(0) = 1$ to obtain the integration constant C.

$$y(0) = 2e^{(2)(0)}(0 - 1) + Ce^0 = 1$$
$$(2)(1)(-1) + C(1) = 1$$
$$-2 + C = 1$$
$$C = 3$$

Substituting in the value for the integration constant, C, the solution is

$$y = \boxed{2e^{2x}(x-1) + 3e^x}$$

The answer is (B).

3. (a) The differential equation is a homogeneous second-order linear differential equation with constant coefficients. Write the characteristic equation.

$$r^2 + 2r + 2 = 0$$

This is a quadratic equation of the form $ar^2 + br + c = 0$, where a equals 1, b equals 2, and c equals 2.

Solve for r.

$$r = \frac{-b \pm \sqrt{b^2 - 4ac}}{2a}$$

$$= \frac{-2 \pm \sqrt{(2)^2 - (4)(1)(2)}}{(2)(1)}$$

$$= \frac{-2 \pm \sqrt{4 - 8}}{2}$$

$$= \frac{-2 \pm \sqrt{-4}}{2}$$

$$= \frac{-2 \pm 2\sqrt{-1}}{2}$$

$$= -1 \pm \sqrt{-1}$$

$$= -1 \pm i$$

$$r_1 = -1 + i \text{ and } r_2 = -1 - i$$

Because the roots are imaginary and of the form $\alpha + i\omega$ and $\alpha - i\omega$, where α equals -1 and ω equals 1, the general form of the solution is given by

$$x(t) = A_1 e^{\alpha t} \cos \omega t + A_2 e^{\alpha t} \sin \omega t$$

$$= A_1 e^{-1t} \cos(1t) + A_2 e^{-1t} \sin(1t)$$

$$= A_1 e^{-t} \cos t + A_2 e^{-t} \sin t$$

Apply the initial conditions $x(0) = 0$ and $x'(0) = 1$ to solve for A_1 and A_2.

First, apply the initial condition $x(0) = 0$.

$$x(t) = A_1 e^0 \cos 0 + A_2 e^0 \sin 0 = 0$$

$$A_1(1)(1) + A_2(1)(0) = 0$$

$$A_1 = 0$$

Substituting, the solution of the differential equation becomes

$$x(t) = A_2 e^{-t} \sin t$$

To apply the second initial condition, take the first derivative.

$$x'(t) = \frac{d}{dt}\left(A_2 e^{-t} \sin t\right)$$

$$= A_2 \frac{d}{dt}\left(e^{-t} \sin t\right)$$

$$= A_2\left(\sin t \frac{d}{dt}(e^{-t}) + e^{-t} \frac{d}{dt} \sin t\right)$$

$$= A_2\left(\sin t(-e^{-t}) + e^{-t} \cos t\right)$$

$$= A_2(e^{-t})(-\sin t + \cos t)$$

Apply the initial condition, $x'(0) = 1$.

$$x(0) = A_2(e^0)(-\sin 0 + \cos 0) = 1$$

$$A_2(1)(0 + 1) = 1$$

$$A_2 = 1$$

The solution is

$$x(t) = A_2 e^{-t} \sin t$$

$$= (1)e^{-t} \sin t$$

$$= \boxed{e^{-t} \sin t}$$

The answer is (C).

(b) To determine the natural frequency, set the damping term to zero. The equation has the form

$$x'' + 2x = 0$$

This equation has a general solution of the form

$$x(t) = x_0 \cos \omega t + \left(\frac{v_0}{\omega}\right) \sin \omega t$$

ω is the natural frequency.

Given the equation $x'' + 2x = 0$, the characteristic equation is

$$r^2 + 2 = 0$$

$$r = \sqrt{-2}$$

$$= \pm\sqrt{2}i$$

Because the roots are imaginary and of the form $\alpha + i\omega$ and $\alpha - i\omega$, where α equals 0 and ω equals $\sqrt{2}$, the general form of the solution is given by

$$x(t) = A_1 e^{\alpha t} \cos \omega t + A_2 e^{\alpha t} \sin \omega t$$

$$= A_1 e^{0t} \cos \sqrt{2}t + A_2 e^{0t} \sin \sqrt{2}t$$

$$= A_1(1)\cos \sqrt{2}t + A_2(1)\sin \sqrt{2}t$$

$$= A_1 \cos \sqrt{2}t + A_2 \sin \sqrt{2}t$$

Apply the initial conditions $x(0) = 0$ and $x'(0) = 1$ to solve for A_1 and A_2. Applying the initial condition $x(0) = 0$ gives

$$x(0) = A_1 \cos\big((\sqrt{2})(0)\big) + A_2 \sin\big((\sqrt{2})(0)\big) = 0$$
$$A_1 \cos 0 + A_2 \sin 0 = 0$$
$$A_1(1) + A_2(0) = 0$$
$$A_1 = 0$$

Substituting, the solution of the different equation becomes

$$x(t) = A_2 \sin \sqrt{2}t$$

To apply the second initial condition, take the first derivative.

$$x'(t) = \frac{d}{dt}(A_2 \sin \sqrt{2}t)$$
$$= A_2\sqrt{2} \cos \sqrt{2}t$$

Apply the second initial condition, $x'(0) = 1$.

$$x'(0) = A_2\sqrt{2} \cos\big((\sqrt{2})(0)\big) = 1$$
$$A_2\sqrt{2} \cos 0 = 1$$
$$A_2(\sqrt{2})(1) = 1$$
$$A_2\sqrt{2} = 1$$
$$A_2 = \frac{1}{\sqrt{2}} = \frac{\sqrt{2}}{2}$$

Substituting, the undamped solution becomes

$$x(t) = \frac{\sqrt{2}}{2} \sin \sqrt{2}t$$

Therefore, the undamped natural frequency is,

$$\boxed{\omega = \sqrt{2}}$$

The answer is (C).

(c) The amplitude of the oscillation is the maximum displacement.

Take the derivative of the solution, $x(t) = e^{-t} \sin t$.

$$x'(t) = \frac{d}{dt} e^{-t} \sin t$$
$$= (\sin t)\frac{d}{dt}(e^{-t}) + (e^{-t})\frac{d}{dt} \sin t$$
$$= \sin t(-e^{-t}) + (e^{-t})\cos t$$
$$= e^{-t}(\cos t - \sin t)$$

The maximum displacement occurs at $x'(t) = 0$.

Because $e^{-t} \neq 0$ except as t approaches infinity,

$$\cos t - \sin t = 0$$
$$\tan t = 1$$
$$t = \tan^{-1} 1$$
$$= 0.785 \text{ rad}$$

At time $t = 0.785$ rad, the displacement is at a maximum. Substitute into the original solution to obtain a value for the maximum displacement.

$$x(0.785) = e^{-0.785} \sin 0.785$$
$$= 0.32$$

The amplitude is $\boxed{0.32.}$

The answer is (A).

(d) (An alternative solution using Laplace transforms follows this solution.) The application of a lateral wind load with the form $\sin t$ revises the differential equation to the form

$$x'' + 2x' + 2x = \sin t$$

Express the solution as the sum of the complementary x_c and particular x_p solutions.

$$x(t) = x_c(t) + x_p(t)$$

From part (a),

$$x_c(t) = A_1 e^{-t} \cos t + A_2 e^{-t} \sin t$$

The general form of the particular solution is given by

$$x_p(t) = x^s(A_3 \cos t + A_4 \sin t)$$

Determine the value of s; check to see if the terms of the particular solution solve the homogeneous equation.

Examine the term $A_3 \cos t$.

Take the first derivative.

$$\frac{d}{dt}(A_3 \cos t) = -A_3 \sin t$$

Take the second derivative.

$$\frac{d}{dt}\left(\frac{d}{dt}(A_3 \cos t)\right) = \frac{d}{dt}(-A_3 \sin t)$$
$$= -A_3 \cos t$$

Substitute the terms into the homogeneous equation.

$$x'' + 2x' + 2x = -A_3 \cos t + 2(-A_3 \sin t)$$
$$+ 2(-A_3 \cos t)$$
$$= A_3 \cos t - 2A_3 \sin t$$
$$\neq 0$$

Except for the trivial solution $A_3 = 0$, the term $A_3 \cos t$ does not solve the homogeneous equation.

Examine the second term $A_4 \sin t$.

Take the first derivative.

$$\frac{d}{dt}(A_4 \sin t) = A_4 \cos t$$

Take the second derivative.

$$\frac{d}{dt}\left(\frac{d}{dt}(A_4 \sin t)\right) = \frac{d}{dt}(A_4 \cos t)$$
$$= -A_4 \sin t$$

Substitute the terms into the homogeneous equation.

$$x'' + 2x' + 2x = -A_4 \sin t + 2(A_4 \cos t)$$
$$+ 2(A_4 \sin t)$$
$$= A_4 \sin t + 2A_4 \cos t$$
$$\neq 0$$

Except for the trivial solution $A_4 = 0$, the term $A_4 \sin t$ does not solve the homogeneous equation.

Neither of the terms satisfies the homogeneous equation $s = 0$; therefore, the particular solution is of the form

$$x_p(t) = A_3 \cos t + A_4 \sin t$$

Use the method of undetermined coefficients to solve for A_3 and A_4. Take the first derivative.

$$x_p'(t) = \frac{d}{dt}(A_3 \cos t + A_4 \sin t)$$
$$= -A_3 \sin t + A_4 \cos t$$

Take the second derivative.

$$x_p''(t) = \frac{d}{dt}\left(\frac{d}{dt}(A_3 \cos t + A_4 \sin t)\right)$$
$$= \frac{d}{dt}(-A_3 \sin t + A_4 \cos t)$$
$$= -A_3 \cos t - A_4 \sin t$$

Substitute the expressions for the derivatives into the differential equation.

$$x'' + 2x' + 2x = (-A_3 \cos t - A_4 \sin t)$$
$$+ 2(-A_3 \sin t + A_4 \cos t)$$
$$+ 2(A_3 \cos t + A_4 \sin t)$$
$$= \sin t$$

Rearranging terms gives

$$(-A_3 + 2A_4 + 2A_3)\cos t$$
$$+ (-A_4 - 2A_3 + 2A_4)\sin t = \sin t$$
$$(A_3 + 2A_4)\cos t + (-2A_3 + A_4)\sin t = \sin t$$

Equating coefficients gives

$$A_3 + 2A_4 = 0$$
$$-2A_3 + A_4 = 1$$

Multiplying the first equation by 2 and adding equations gives

$$2A_3 + 4A_4 = 0$$
$$+(-2A_3 + A_4) = 1$$
$$\overline{\qquad 5A_4 = 1 \text{ or } A_4 = \tfrac{1}{5}}$$

From the first equation for $A_4 = 1/5$, $A_3 + (2)(1/5) = 0$ and $A_3 = -2/5$.

Substituting for the coefficients, the particular solution becomes

$$x_p(t) = -\tfrac{2}{5}\cos t + \tfrac{1}{5}\sin t$$

Combining the complementary and particular solutions gives

$$x(t) = x_c(t) + x_p(t)$$
$$= A_1 e^{-t}\cos t + A_2 e^{-t}\sin t - \tfrac{2}{5}\cos t + \tfrac{1}{5}\sin t$$

Apply the initial conditions to solve for the coefficients A_1 and A_2; then apply the first initial condition, $x(0) = 0$.

$$x(t) = A_1 e^0 \cos 0 + A_2 e^0 \sin 0$$
$$- \tfrac{2}{5}\cos 0 + \tfrac{1}{5}\sin 0 = 0$$
$$A_1(1)(1) + A_2(1)(0) + (-\tfrac{2}{5})(1) + (\tfrac{1}{5})(0) = 0$$
$$A_1 - \tfrac{2}{5} = 0$$
$$A_1 = \tfrac{2}{5}$$

Substituting for A_1, the solution becomes

$$x(t) = \tfrac{2}{5}e^{-t}\cos t + A_2 e^{-t}\sin t - \tfrac{2}{5}\cos t + \tfrac{1}{5}\sin t$$

Take the first derivative.

$$x'(t) = \frac{d}{dt}\left(\tfrac{2}{5}e^{-t}\cos t + A_2 e^{-t}\sin t\right)$$
$$+ \left((-\tfrac{2}{5})\cos t + \tfrac{1}{5}\sin t\right)$$
$$= (\tfrac{2}{5})(-e^{-t}\cos t - e^{-t}\sin t)$$
$$+ A_2(-e^{-t}\sin t + e^{-t}\cos t)$$
$$+ (-\tfrac{2}{5})(-\sin t) + \tfrac{1}{5}\cos t$$

Apply the second initial condition, $x'(0)=1$.

$$x'(0) = \frac{2}{5}(-e^0 \cos 0 - e^0 \sin 0)$$
$$+ A_2(-e^0 \sin 0 + e^0 \cos 0)$$
$$+ \left(-\frac{2}{5}\right)(-\sin 0) + \frac{1}{5}\cos 0$$
$$= 1$$

$$\left(\frac{2}{5}\right)\left(-(1)(1) - (1)(0)\right)$$
$$+ A_2\left(-(1)(0) + (1)(1)\right) + \left(-\frac{2}{5}\right)(0) + \left(\frac{1}{5}\right)(1) = 1$$
$$\left(\frac{2}{5}\right)(-1) + A_2(1) + \left(\frac{1}{5}\right) = 1$$
$$A_2 = \frac{6}{5}$$

Substituting for A_2, the solution becomes

$$x(t) = \boxed{\begin{array}{l} \frac{2}{5}e^{-t}\cos t + \frac{6}{5}e^{-t}\sin t \\ -\frac{2}{5}\cos t + \frac{1}{5}\sin t \end{array}}$$

The answer is (D).

(d) *Alternate Solution:*

Use the Laplace transform method.

$$x'' + 2x' + 2x = \sin t$$
$$\mathcal{L}(x'') + 2\mathcal{L}(x') + 2\mathcal{L}(x) = \mathcal{L}\sin t$$
$$s^2\mathcal{L}(x) - 1 + 2s\mathcal{L}(x) + 2\mathcal{L}(x) = \frac{1}{s^2+1}$$
$$\mathcal{L}(x)(s^2 + 2s + 2) - 1 = \frac{1}{s^2+1}$$
$$\mathcal{L}(x) = \frac{1}{s^2+2s+2} + \frac{1}{(s^2+1)(s^2+2s+2)}$$
$$= \frac{1}{(s+1)^2+1} + \frac{1}{(s^2+1)(s^2+2s+2)}$$

Use partial fractions to expand the second term.

$$\frac{1}{(s^2+1)(s^2+2s+2)} = \frac{A_1+B_1 s}{s^2+1} + \frac{A_2+B_2 s}{s^2+2s+2}$$

Cross multiply.

$$\frac{A_1+B_1 s}{s^2+1} + \frac{A_2+B_2 s}{s^2+2s+2}$$
$$= \frac{A_1 s^2 + 2A_1 s + 2A_1 + B_1 s^3 + 2B_1 s^2}{(s^2+1)(s^2+2s+2)}$$
$$\qquad\qquad \frac{+ 2B_1 s + A_2 s^2 + A_2 + B_2 s^3 + B_2 s}{}$$
$$= \frac{s^3(B_1+B_2) + s^2(A_1+A_2+2B_1)}{(s^2+1)(s^2+2s+2)}$$
$$\qquad\qquad \frac{+ s(2A_1+2B_1+B_2) + 2A_1 + A_2}{}$$

$$B_1 + B_2 = 0$$
$$A_1 + A_2 + 2B_1 = 0$$
$$2A_1 + 2B_1 + B_2 = 0$$
$$2A_1 + A_2 = 1$$

Use Cramer's rule to find A_1.

$$A_1 = \frac{\begin{vmatrix} 0 & 0 & 1 & 1 \\ 0 & 1 & 2 & 0 \\ 0 & 0 & 2 & 1 \\ 1 & 1 & 0 & 0 \end{vmatrix}}{\begin{vmatrix} 0 & 0 & 1 & 1 \\ 1 & 1 & 2 & 0 \\ 2 & 0 & 2 & 1 \\ 2 & 1 & 0 & 0 \end{vmatrix}} = \frac{-1}{-5} = \frac{1}{5}$$

The rest of the coefficients are found similarly.

$$A_1 = \frac{1}{5}$$
$$A_2 = \frac{3}{5}$$
$$B_1 = -\frac{2}{5}$$
$$B_2 = \frac{2}{5}$$

Then,

$$\mathcal{L}(x) = \frac{1}{(s+1)^2+1} + \frac{\frac{1}{5}}{s^2+1} + \frac{-\frac{2}{5}s}{s^2+1}$$
$$+ \frac{\frac{3}{5}}{s^2+2s+2} + \frac{\frac{2}{5}s}{s^2+2s+2}$$

Take the inverse transform.

$$x(t) = \mathcal{L}^{-1}\{\mathcal{L}(x)\}$$
$$= e^{-t}\sin t + \frac{1}{5}\sin t - \frac{2}{5}\cos t + \frac{3}{5}e^{-t}\sin t$$
$$+ \frac{2}{5}(e^{-t}\cos t - e^{-t}\sin t)$$
$$= \boxed{\frac{2}{5}e^{-t}\cos t + \frac{6}{5}e^{-t}\sin t - \frac{2}{5}\cos t + \frac{1}{5}\sin t}$$

The answer is (D).

4. *Customary U.S. Solution*

The differential equation is given as

$$m'(t) = a(t) = \frac{m(t)o(t)}{V(t)}$$

$a(t) =$ rate of addition of chemical

$m(t) =$ mass of chemical at time t

$o(t) =$ volumetric flow out of the lagoon (30 gal/min)

$V(t) =$ volume in the lagoon at time t

$d =$ diameter of the lagoon

$z =$ depth of the lagoon

Water flows into the lagoon at a rate of 30 gal/min, and a water-chemical mix flows out of the lagoon at a rate of 30 gal/min. Therefore, the volume of the lagoon at time t is equal to the initial volume.

$$V(t) = \frac{\pi}{4} d^2 z$$

$$= \left(\frac{\pi}{4}\right)(120 \text{ ft})^2 (10 \text{ ft})$$

$$= 113{,}097 \text{ ft}^3$$

Use a conversion factor of 7.48 gal/ft^3.

$$o(t) = \frac{30 \frac{\text{gal}}{\text{min}}}{7.48 \frac{\text{gal}}{\text{ft}^3}}$$

$$= 4.01 \text{ ft}^3/\text{min}$$

Substituting into the general form of the differential equation gives

$$m'(t) = a(t) - \frac{m(t)\, o(t)}{V(t)}$$

$$= 0 \text{ lbm} - m(t) \left(\frac{4.01 \frac{\text{ft}^3}{\text{min}}}{113{,}097 \text{ ft}^3}\right)$$

$$= -\left(\frac{3.55 \times 10^{-5}}{\text{min}}\right) m(t)$$

$$m'(t) + \left(\frac{3.55 \times 10^{-5}}{\text{min}}\right) m(t)$$

$$= 0$$

The differential equation of the problem has a characteristic equation.

$$r + \frac{3.55 \times 10^{-5}}{\text{min}} = 0$$

$$r = -3.55 \times 10^{-5}/\text{min}$$

The general form of the solution is

$$m(t) = A e^{rt}$$

Substituting for the root, r, gives

$$m(t) = A e^{(-3.55 \times 10^{-5}/\text{min})t}$$

Apply the initial condition $m(0) = 90$ lbm at time $t = 0$ min.

$$m(0) = A e^{(-3.55 \times 10^{-5}/\text{min})(0)} = 90 \text{ lbm}$$

$$A e^0 = 90 \text{ lbm}$$

$$A = 90 \text{ lbm}$$

Therefore,

$$m(t) = (90 \text{ lbm}) e^{(-3.55 \times 10^{-5}/\text{min})t}$$

Solve for t.

$$\frac{m(t)}{90 \text{ lbm}} = e^{(-3.55 \times 10^{-5}/\text{min})t}$$

$$\ln \frac{m(t)}{90 \text{ lbm}} = \ln e^{(-3.55 \times 10^{-5}/\text{min})t}$$

$$= \frac{-3.55 \times 10^{-5}}{\text{min}} t$$

$$t = \frac{\ln \frac{m(t)}{90 \text{ lbm}}}{\frac{-3.55 \times 10^{-5}}{\text{min}}}$$

The initial mass of the water in the lagoon is

$$m_i = V \rho$$

$$= (113{,}097 \text{ ft}^3)\left(62.4 \frac{\text{lbm}}{\text{ft}^3}\right)$$

$$= 7.06 \times 10^6 \text{ lbm}$$

The final mass of chemicals is achieved at a concentration of 1 ppb or

$$m_f = \frac{m_i}{1 \times 10^9} = \frac{7.06 \times 10^6 \text{ lbm}}{1 \times 10^9}$$

$$= 7.06 \times 10^{-3} \text{ lbm}$$

Find the time required to achieve a mass of 7.06×10^{-3} lbm.

$$t = \left(\frac{\ln \frac{m(t)}{90 \text{ lbm}}}{\frac{-3.55 \times 10^{-5}}{\text{min}}}\right)\left(\frac{1 \text{ hr}}{60 \text{ min}}\right)\left(\frac{1 \text{ day}}{24 \text{ hr}}\right)$$

$$= \left(\frac{\ln \frac{7.06 \times 10^{-3} \text{ lbm}}{90 \text{ lbm}}}{\frac{-3.55 \times 10^{-5}}{\text{min}}}\right)\left(\frac{1 \text{ hr}}{60 \text{ min}}\right)\left(\frac{1 \text{ day}}{24 \text{ hr}}\right)$$

$$= \boxed{185 \text{ days} \quad (200 \text{ days})}$$

The answer is (D).

SI Solution

The differential equation is given as

$$m'(t) = a(t) - \frac{m(t)o(t)}{V(t)}$$

$a(t) = $ rate of addition of chemical

$m(t) = $ mass of chemical at time t

$o(t) = $ volumetric flow out of the lagoon

(115 L/min)

$V(t) = $ volume in the lagoon at time t

$d = $ diameter of the lagoon

$z = $ depth of the lagoon

Water flows into the lagoon at a rate of 115 L/min, and a water-chemical mix flows out of the lagoon at a rate of 115 L/min. Therefore, the volume of the lagoon at time t is equal to the initial volume.

$$
\begin{aligned}
V(t) &= \frac{\pi}{4}d^2 z \\
&= \left(\frac{\pi}{4}\right)(35\text{ m})^2(3\text{ m}) \\
&= 2886\text{ m}^3
\end{aligned}
$$

Using a conversion factor of 1 m^3/1000 L gives

$$
\begin{aligned}
o(t) &= \left(115\ \frac{\text{L}}{\text{min}}\right)\left(\frac{1\text{ m}^3}{1000\text{ L}}\right) \\
&= 0.115\text{ m}^3/\text{min}
\end{aligned}
$$

Substitute into the general form of the differential equation.

$$
\begin{aligned}
m'(t) &= a(t) - \frac{m(t)o(t)}{V(t)} \\
&= 0\text{ lbm} - m(t)\left(\frac{0.115\ \frac{\text{m}^3}{\text{min}}}{2886\text{ m}^3}\right) \\
&= -\left(\frac{3.985 \times 10^{-5}}{\text{min}}\right)m(t) \\
m'(t) + \left(\frac{3.985 \times 10^{-5}}{\text{min}}\right)m(t) &= 0
\end{aligned}
$$

The differential equation of the problem has the following characteristic equation.

$$
\begin{aligned}
r + \frac{3.985 \times 10^{-5}}{\text{min}} &= 0 \\
r &= -3.985 \times 10^{-5}/\text{min}
\end{aligned}
$$

The general form of the solution is

$$m(t) = Ae^{rt}$$

Substituting in for the root, r, gives

$$m(t) = Ae^{(-3.985\times 10^{-5}/\text{min})t}$$

Apply the initial condition $m(0) = 40$ kg at time $t = 0$ min.

$$
\begin{aligned}
m(0) = Ae^{(-3.985\times 10^{-5}/\text{min})(0)} &= 40\text{ kg} \\
Ae^0 &= 40\text{ kg} \\
A &= 40\text{ kg}
\end{aligned}
$$

Therefore,

$$m(t) = (40\text{ kg})e^{(-3.985\times 10^{-5}/\text{min})t}$$

Solve for t.

$$
\begin{aligned}
\frac{m(t)}{40\text{ kg}} &= e^{(-3.985\times 10^{-5}/\text{min})t} \\
\ln\left(\frac{m(t)}{40\text{ kg}}\right) &= \ln e^{(-3.985\times 10^{-5}/\text{min})t} \\
&= \left(\frac{-3.985\times 10^{-5}}{\text{min}}\right)t \\
t &= \ln\frac{\frac{m(t)}{40\text{ kg}}}{\frac{-3.985\times 10^{-5}}{\text{min}}}
\end{aligned}
$$

The initial mass of water in the lagoon is

$$
\begin{aligned}
m_i &= V\rho \\
&= (2886\text{ m}^3)\left(1000\ \frac{\text{kg}}{\text{m}^3}\right) \\
&= 2.866 \times 10^6\text{ kg}
\end{aligned}
$$

The final mass of chemicals is achieved at a concentration of 1 ppb or

$$
\begin{aligned}
m_f &= \frac{2.886 \times 10^6\text{ kg}}{1 \times 10^9} \\
&= 2.886 \times 10^{-3}\text{ kg}
\end{aligned}
$$

Find the time required to achieve a mass of 2.886×10^{-3} kg.

$$t = \left(\frac{\ln \dfrac{m(t)}{40 \text{ kg}}}{\dfrac{-3.985 \times 10^{-5}}{\text{min}}} \right) \left(\frac{1 \text{ h}}{60 \text{ min}} \right) \left(\frac{1 \text{ d}}{24 \text{ h}} \right)$$

$$= \left(\frac{\ln \left(\dfrac{2.886 \times 10^{-3} \text{ kg}}{40 \text{ kg}} \right)}{\dfrac{-3.985 \times 10^{-5}}{\text{min}}} \right) \left(\frac{1 \text{ h}}{60 \text{ min}} \right) \left(\frac{1 \text{ d}}{24 \text{ h}} \right)$$

$$= \boxed{166 \text{ d} \quad (200 \text{ d})}$$

The answer is (D).

5. Let

$m(t) =$ mass of salt in tank at time t

$m_0 = 60$ mass units

$m'(t) =$ rate at which salt content is changing

Two mass units of salt enter each minute, and three volumes leave each minute. The amount of salt leaving each minute is

$$\left(3 \frac{\text{vol}}{\text{min}} \right) \left(\text{concentration in } \frac{\text{mass}}{\text{vol}} \right)$$

$$= \left(3 \frac{\text{vol}}{\text{min}} \right) \left(\frac{\text{salt content}}{\text{volume}} \right)$$

$$= \left(3 \frac{\text{vol}}{\text{min}} \right) \left(\frac{m(t)}{100 - t} \right)$$

$$m'(t) = 2 - (3) \left(\frac{m(t)}{100 - t} \right)$$

Alternately,

$$m'(t) + \frac{3m(t)}{100 - t} = 2 \text{ mass/min}$$

This is a first-order linear differential equation. The integrating factor is

$$m = e^{3 \int \frac{dt}{100 - t}}$$

$$= e^{(3)\left(-\ln(100 - t) \right)}$$

$$= (100 - t)^{-3}$$

$$m(t) = (100 - t)^3 \left(2 \int \frac{dt}{(100 - t)^3} + k \right)$$

$$= 100 - t + k(100 - t)^3$$

But $m = 60$ mass units at time $t = 0$ min, so k is -0.00004.

$$m(t) = 100 - t - (0.00004)(100 - t)^3$$

At time $t = 60$ min,

$$m = 100 - 60 \text{ min} - (0.00004)(100 - 60 \text{ min})^3$$

$$= \boxed{37.44 \text{ mass units} \quad (37 \text{ mass units})}$$

The answer is (B).

12 Probability and Statistical Analysis of Data

PRACTICE PROBLEMS

1. Four military recruits whose respective shoe sizes are 7, 8, 9, and 10 report to the supply clerk to be issued boots. The supply clerk selects one pair of boots in each of the four required sizes and hands them at random to the recruits.

(a) What is the probability that all recruits will receive boots of an incorrect size?

(A) 0.25

(B) 0.38

(C) 0.45

(D) 0.61

(b) What is the probability that exactly three recruits will receive boots of the correct size?

(A) 0.0

(B) 0.063

(C) 0.17

(D) 0.25

2. The time taken by a toll taker to collect the toll from vehicles crossing a bridge is an exponential distribution with a mean of 23 sec. What is the probability that a random vehicle will be processed in 25 sec or more (i.e., will take longer than 25 sec)?

(A) 0.17

(B) 0.25

(C) 0.34

(D) 0.52

3. The number of cars entering a toll plaza on a bridge during the hour after midnight follows a Poisson distribution with a mean of 20.

(a) What is the probability that 17 cars will pass through the toll plaza during that hour on any given night?

(A) 0.076

(B) 0.12

(C) 0.16

(D) 0.23

(b) What is the probability that three or fewer cars will pass through the toll plaza at that hour on any given night?

(A) 0.0000032

(B) 0.0019

(C) 0.079

(D) 0.11

4. A mechanical component exhibits a negative exponential failure distribution with a mean time to failure, MTTF, of 1000 hr. What is the maximum operating time such that the reliability remains above 99%?

(A) 3.3 hr

(B) 5.6 hr

(C) 8.1 hr

(D) 10 hr

5. A survey field crew measures one leg of a traverse four times. The following results are obtained.

repetition	measurement	direction
1	1249.529	forward
2	1249.494	backward
3	1249.384	forward
4	1249.348	backward

The crew chief is under orders to obtain readings with confidence limits of 90%.

Mathematics

(a) Which readings are acceptable?

  (A) No readings are acceptable.

  (B) Two readings are acceptable.

  (C) Three readings are acceptable.

  (D) All four readings are acceptable.

(b) Which readings are NOT acceptable?

  (A) No readings are unacceptable.

  (B) One reading is unacceptable.

  (C) Two readings are unacceptable.

  (D) All four readings are unacceptable.

(c) Explain how to determine which readings are NOT acceptable.

  (A) Readings inside the 90% confidence limits are unacceptable.

  (B) Readings outside the 90% confidence limits are unacceptable.

  (C) Readings outside the upper 90% confidence limit are unacceptable.

  (D) Readings outside the lower 90% confidence limit are unacceptable.

(d) What is the most probable value of the distance?

  (A) 1249.399

  (B) 1249.410

  (C) 1249.439

  (D) 1249.452

(e) What is the error in the most probable value (at 90% confidence)?

  (A) 0.080

  (B) 0.11

  (C) 0.14

  (D) 0.19

(f) If the distance is one side of a square traverse whose sides are all equal, what is the most probable closure error?

  (A) 0.14

  (B) 0.20

  (C) 0.28

  (D) 0.35

(g) What is the probable error of part (f) expressed as a fraction?

  (A) 1:17,600

  (B) 1:14,200

  (C) 1:12,500

  (D) 1:10,900

(h) What is the order of accuracy of the closure?

  (A) first order

  (B) second order

  (C) third order

  (D) fourth order

(i) Define accuracy, and distinguish it from precision.

  (A) If an experiment can be repeated with identical results, the results are considered accurate.

  (B) If an experiment has a small bias, the results are considered precise.

  (C) If an experiment is precise, it cannot also be accurate.

  (D) If an experiment is unaffected by experimental error, the results are accurate.

(j) Give an example of systematic error.

  (A) measuring a river's depth as a motorized ski boat passes by

  (B) using a steel tape that is too short to measure consecutive distances

  (C) locating magnetic north near a large iron ore deposit along an overland route

  (D) determining local wastewater BOD after a toxic spill

6. Prior to the use of radar speed control on a road, California law requires a statistical analysis of the average speed driven there by motorists. The following speeds (all in mph) were observed in a random sample of 40 cars.

44, 48, 26, 25, 20, 43, 40, 42, 29, 39, 23, 26, 24, 47, 45, 28, 29, 41, 38, 36, 27, 44, 42, 43, 29, 37, 34, 31, 33, 30, 42, 43, 28, 41, 29, 36, 35, 30, 32, 31

(a) Tabulate the frequency distribution of the data.

(b) Draw the frequency histogram.

(c) Draw the frequency polygon.

(d) Tabulate the cumulative frequency distribution.

(e) Draw the cumulative frequency graph.

(f) What is the upper quartile speed?

(A) 30 mph

(B) 35 mph

(C) 40 mph

(D) 45 mph

(g) What is the median speed?

(A) 31 mph

(B) 33 mph

(C) 35 mph

(D) 37 mph

(h) What is the standard deviation of the sample data?

(A) 2.1 mph

(B) 6.1 mph

(C) 6.8 mph

(D) 7.4 mph

(i) What is the sample standard deviation?

(A) 7.5 mph

(B) 18 mph

(C) 35 mph

(D) 56 mph

(j) What is most nearly the sample variance?

(A) 56 $\mathrm{mi^2/hr^2}$

(B) 320 $\mathrm{mi^2/hr^2}$

(C) 1230 $\mathrm{mi^2/hr^2}$

(D) 3140 $\mathrm{mi^2/hr^2}$

7. A spot speed study is conducted for a stretch of roadway. During a normal day, the speeds were found to be normally distributed with a mean of 46 mph and a standard deviation of 3 mph.

(a) What is the 50th percentile speed?

(A) 39 mph

(B) 43 mph

(C) 46 mph

(D) 49 mph

(b) What is the 85th percentile speed?

(A) 47.1 mph

(B) 48.3 mph

(C) 49.1 mph

(D) 52.7 mph

(c) What speed is two standard deviations above the mean?

(A) 47.2 mph

(B) 49.3 mph

(C) 51.1 mph

(D) 52.0 mph

(d) The daily average speeds for the same stretch of roadway on consecutive normal days were determined by sampling 25 vehicles each day. What speed is two standard deviations above the mean?

(A) 46.6 mph

(B) 47.2 mph

(C) 52.0 mph

(D) 54.7 mph

8. The diameters of bolt holes drilled in structural steel members are normally distributed with a mean of 0.502 in and a standard deviation of 0.005 in. Holes are out of specification if their diameters are less than 0.497 in or more than 0.507 in.

(a) What is the probability that a hole chosen at random will be out of specification?

(A) 0.16

(B) 0.22

(C) 0.32

(D) 0.68

(b) What is the probability that 2 holes out of a sample of 15 will be out of specification?

(A) 0.074

(B) 0.12

(C) 0.15

(D) 0.32

9. 100 bearings were tested to failure. The average life was 1520 hr, and the standard deviation was 120 hr. The manufacturer claims a 1600 hr life. Evaluate using confidence limits of 95% and 99%.

(A) The claim is accurate at both 95% and 99% confidence.

(B) The claim is inaccurate only at 95%.

(C) The claim is inaccurate only at 99%.

(D) The claim is inaccurate at both 95% and 99% confidence.

Mathematics

10. (a) Find the best equation for a line passing through the points given. (b) Find the correlation coefficient.

x	y
400	370
800	780
1250	1210
1600	1560
2000	1980
2500	2450
4000	3950

11. Find the best equation for a line passing through the points given.

s	t
20	43
18	141
16	385
14	1099

12. The number of vehicles lining up behind a flashing railroad crossing has been observed for five trains of different lengths, as given. What is the mathematical formula that relates the two variables?

no. of cars in train	no. of vehicles
2	14.8
5	18.0
8	20.4
12	23.0
27	29.9

13. The following yield data are obtained from five identical treatment plants. (a) Develop a mathematical equation to correlate the yield and average temperature. (b) What is the correlation coefficient?

treatment plant	average temperature (T)	average yield (Y)
1	207.1	92.30
2	210.3	92.58
3	200.4	91.56
4	201.1	91.63
5	203.4	91.83

14. The following data are obtained from a soil compaction test. What is the mathematical formula that relates the two variables?

x	y
−1	0
0	1
1	1.4
2	1.7
3	2
4	2.2
5	2.4
6	2.6
7	2.8
8	3

15. Two resistances, the meter resistance and a shunt resistor, are connected in parallel in an ammeter. Most of the current passing through the meter goes through the shunt resistor. In order to determine the accuracy of the resistance of shunt resistors being manufactured for a line of ammeters, a manufacturer tests a sample of 100 shunt resistors. The numbers of shunt resistors with the resistance indicated (to the nearest hundredth of an ohm) are as follows.

0.200 Ω, 1; 0.210 Ω, 3; 0.220 Ω, 5; 0.230 Ω, 10; 0.240 Ω, 17; 0.250 Ω, 40; 0.260 Ω, 13; 0.270 Ω, 6; 0.280 Ω, 3; 0.290 Ω, 2

(a) What is the mean resistance?

  (A) 0.235 Ω

  (B) 0.247 Ω

  (C) 0.251 Ω

  (D) 0.259 Ω

(b) What is the sample standard deviation?

  (A) 0.00030

  (B) 0.010

  (C) 0.016

  (D) 0.24

(c) What is the median resistance?

  (A) 0.221 Ω

  (B) 0.244 Ω

  (C) 0.252 Ω

  (D) 0.259 Ω

(d) What is most nearly the sample variance?

(A) $0.00027 \ \Omega^2$

(B) $0.0083 \ \Omega^2$

(C) $0.011 \ \Omega^2$

(D) $0.016 \ \Omega^2$

SOLUTIONS

1. (a) There are $4! = 24$ different possible outcomes. By enumeration, the number of completely wrong combinations is 9, so that

$$p\{\text{all wrong}\} = \frac{9}{24} = \boxed{0.375 \quad (0.38)}$$

The answer is (B).

(b) If three recruits get the correct size, the fourth recruit will also, because there will be only one pair remaining.

$$p\{\text{exactly 3}\} = \boxed{0}$$

The answer is (A).

2. For an exponential distribution function, the mean is given as

$$\mu = \frac{1}{\lambda}$$

For a mean of 23,

$$\mu = 23 = \frac{1}{\lambda}$$
$$\lambda = 0.0435$$

For an exponential distribution function,

$$p\{X < x\} = F(x) = 1 - e^{-\lambda x}$$
$$p\{x > X\} = 1 - p\{X < x\}$$
$$= 1 - F(x)$$
$$= 1 - (1 - e^{-\lambda x})$$
$$= e^{-\lambda x}$$

The probability of a random vehicle being processed in 25 sec or more is

$$p\{x > 25\} = e^{-(0.0435)(25)}$$
$$= e^{-1.0875}$$
$$= \boxed{0.337 \quad (0.34)}$$

The answer is (C).

3. (a) The distribution is a Poisson distribution with an average of $\lambda = 20$. The probability for a Poisson distribution is

$$p\{x\} = f(x) = \frac{e^{-\lambda}\lambda^x}{x!}$$

The probability of 17 cars is

$$p\{x = 17\} = f(17) = \frac{e^{-20}(20)^{17}}{17!}$$
$$= \boxed{0.076 \quad (7.6\%)}$$

The answer is (A).

(b) The probability of three or fewer cars is

$$p\{x \leq 3\} = p\{x = 0\} + p\{x = 1\} + p\{x = 2\}$$
$$+ p\{x = 3\}$$
$$= f(0) + f(1) + f(2) + f(3)$$
$$= \frac{e^{-20}(20)^0}{0!} + \frac{e^{-20}(20)^1}{1!}$$
$$+ \frac{e^{-20}(20)^2}{2!} + \frac{e^{-20}(20)^3}{3!}$$
$$= 2 \times 10^{-9} + 4.1 \times 10^{-8}$$
$$+ 4.12 \times 10^{-7} + 2.75 \times 10^{-6}$$
$$= \boxed{3.2 \times 10^{-6} \quad (0.0000032)}$$

The answer is (A).

4. To simplify the notation,

$$\lambda = \frac{1}{\text{MTTF}}$$

$$\frac{1}{1000 \text{ hr}} = 0.001 \text{ hr}^{-1}$$

The reliability function is

$$R\{t\} = e^{-\lambda t/\text{hr}} = e^{-0.001t/\text{hr}}$$

Because the reliability is greater than 99%,

$$e^{-0.001t/\text{hr}} > 0.99$$
$$\ln(e^{-0.001t/\text{hr}}) > \ln 0.99$$
$$\frac{-0.001t}{\text{hr}} > \ln 0.99$$
$$t < (-1000 \text{ hr}) \ln 0.99$$
$$\boxed{t < 10.05 \text{ hr} \quad (10 \text{ hr})}$$

The maximum operating time such that the reliability remains above 99% is 10 hr.

The answer is (D).

5. Find the average.

$$\bar{x} = \frac{\sum x_i}{n}$$
$$= \frac{1249.529 + 1249.494 + 1249.384 + 1249.348}{4}$$
$$= 1249.439$$

Because the sample population is small, use the sample standard deviation.

$$s = \sqrt{\frac{\sum(x_i - \bar{x})^2}{n-1}}$$
$$= \sqrt{\frac{\begin{array}{c}(1249.529 - 1249.439)^2 + (1249.494 - 1249.439)^2\\ + (1249.384 - 1249.439)^2 + (1249.348 - 1249.439)^2\end{array}}{4-1}}$$
$$= 0.08647$$

From the standard deviation table, a 90% confidence limit falls within $1.645s$ of $\bar{x}$.

$$1249.439 \pm (1.645)(0.08647) = 1249.439 \pm 0.142$$

Therefore, (1249.297, 1249.581) is the 90% confidence range.

(a) | By observation, all the readings fall within the 90% confidence range, so all are acceptable. |

The answer is (D).

(b) | No readings are unacceptable. |

The answer is (A).

(c) | Readings outside the 90% confidence limits are unacceptable. |

The answer is (B).

(d) The unbiased estimate of the most probable distance is | 1249.439. |

The answer is (C).

(e) The error for the 90% confidence range is | 0.14. |

The answer is (C).

(f) If the surveying crew places a marker, measures a distance x, places a second marker, and then measures the same distance x back to the original marker, the ending point should coincide with the original marker. If, due to measurement errors, the ending and starting points do not coincide, the difference is the closure error.

In this example, the survey crew moves around the four sides of a square, so there are two measurements in the x-direction and two measurements in the y-direction. If the errors E_1 and E_2 are known for two measurements, x_1 and x_2, the error associated with the sum or difference $x_1 \pm x_2$ is

$$E\{x_1 \pm x_2\} = \sqrt{E_1^2 + E_2^2}$$

In this case, the error in the x-direction is

$$E_x = \sqrt{(0.1422)^2 + (0.1422)^2}$$
$$= 0.2011$$

The error in the y-direction is calculated the same way and is also 0.2011. E_x and E_y are combined by the Pythagorean theorem to yield

$$E_{\text{closure}} = \sqrt{(0.2011)^2 + (0.2011)^2}$$
$$= \boxed{0.2844 \quad (0.28)}$$

The answer is (C).

(g) In surveying, error may be expressed as a fraction of one or more legs of the traverse. Assume that the total of all four legs is to be used as the basis.

$$E = \frac{E_{\text{closure}}}{\sum_i x_i} = \frac{0.2844}{(4)(1249)} = \boxed{\frac{1}{17,567} \quad (1:17,600)}$$

The answer is (A).

(h) In surveying, a class 1 third-order error is smaller than 1/10,000. The error of 1/17,567 is smaller than the third-order error; therefore, the error is within the third-order accuracy.

The answer is (C).

(i) An experiment is accurate if it is unchanged by experimental error. Precision is concerned with the repeatability of the experimental results. If an experiment is repeated with identical results, the experiment is said to be precise. However, it is possible to have a highly precise experiment with a large bias.

The answer is (D).

(j) A systematic error is one that is always present and is unchanged from sample to sample. For example, a steel tape that is 0.02 ft short introduces a systematic error.

The answer is (B).

6. For (a) and (d), tabulate the frequency distribution data.

(The lowest speed is 20 mph and the highest speed is 48 mph; therefore, the range is 28 mph. Choose 10 cells with a width of 3 mph.)

midpoint	interval (mph)	frequency	cumulative frequency	cumulative percent
21	20–22	1	1	3
24	23–25	3	4	10
27	26–28	5	9	23
30	29–31	8	17	43
33	32–34	3	20	50
36	35–37	4	24	60
39	38–40	3	27	68
42	41–43	8	35	88
45	44–46	3	38	95
48	47–49	2	40	100

(b)

(c)

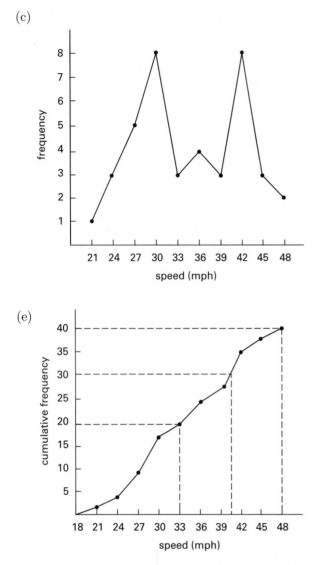

(e)

(f) From the cumulative frequency graph in part (e), the upper quartile speed occurs at 30 cars or 75%, which corresponds to approximately 40 mph.

The answer is (C).

(g) The mode occurs at two speeds (the frequency at each of the two speeds is 8): 30 mph and 42 mph.

The median occurs at 50% (or 20 cars) and, from the cumulative frequency chart, corresponds to 33 mph.

$$\sum x_i = 1390 \text{ mi/hr}$$

$$n = 40$$

The answer is (B).

(h) The standard deviation of the sample data is

$$\sigma = \sqrt{\frac{\sum x^2}{n} - \mu^2}$$

$$\sum x^2 = 50{,}496 \text{ mi}^2/\text{hr}^2$$

The mean is computed as

$$\overline{x} = \frac{\sum x_i}{n}$$

$$= \frac{1390 \ \frac{\text{mi}}{\text{hr}}}{40}$$

$$= \boxed{34.75 \text{ mph} \quad (35 \text{ mph})}$$

Use the sample mean as an unbiased estimator of the population mean, μ.

$$\sigma = \sqrt{\frac{\sum x^2}{n} - \mu^2}$$

$$= \sqrt{\frac{50{,}496 \ \frac{\text{mi}^2}{\text{hr}^2}}{40} - \left(34.75 \ \frac{\text{mi}}{\text{hr}}\right)^2}$$

$$= \boxed{7.405 \text{ mph} \quad (7.4 \text{ mph})}$$

The answer is (D).

(i) The sample standard deviation is

$$s = \sqrt{\frac{\sum x^2 - \frac{\left(\sum x\right)^2}{n}}{n-1}}$$

$$= \sqrt{\frac{50{,}496 \ \frac{\text{mi}^2}{\text{hr}^2} - \frac{\left(1390 \ \frac{\text{mi}}{\text{hr}}\right)^2}{40}}{40-1}}$$

$$= \boxed{7.500 \text{ mph} \quad (7.5 \text{ mph})}$$

The answer is (A).

(j) The sample variance is given by the square of the sample standard deviation.

$$s^2 = \left(7.500 \ \frac{\text{mi}}{\text{hr}}\right)^2$$

$$= \boxed{56.25 \text{ mi}^2/\text{hr}^2 \quad (56 \text{ mi}^2/\text{hr}^2)}$$

The answer is (A).

7. (a) The 50th percentile speed is the mean speed,

$$\boxed{46 \text{ mph}}$$

The answer is (C).

(b) The 85th percentile speed is the speed that is exceeded by only 15% of the measurements. Because this is a normal distribution, App. 12.A can be used. 15% in the upper tail corresponds to 35% between the mean and the 85th percentile. This occurs at approximately 1.04σ. The 85th percentile speed is

$$x_{85\%} = \mu + 1.04\sigma$$
$$= 46 \text{ mph} + (1.04)(3 \text{ mph})$$
$$= \boxed{49.12 \text{ mph} \quad (49.1 \text{ mph})}$$

The answer is (C).

(c) The upper two standard deviation speed, 2σ, is

$$x_{2\sigma} = \mu + 2\sigma$$
$$= 46 \text{ mph} + (2)(3 \text{ mph}) = \boxed{52 \text{ mph}}$$

The answer is (D).

(d) According to the central limit theorem, the mean of the average speeds is the same as the distribution mean, and, for a sample size of K measurements, the standard deviation of sample means is

$$\sigma_{\bar{x}} = \frac{\sigma_x}{\sqrt{K}}$$
$$= \frac{3 \text{ mph}}{\sqrt{25}} = 0.6 \text{ mph}$$
$$\bar{x}_{2\sigma} = \mu + 2\sigma_{\bar{x}}$$
$$= 46 \text{ mph} + (2)(0.6 \text{ mph}) = \boxed{47.2 \text{ mph}}$$

The answer is (B).

8. (a) From Eq. 12.43,

$$z = \frac{x_o - \mu}{\sigma}$$
$$z_{\text{upper}} = \frac{0.507 \text{ in} - 0.502 \text{ in}}{0.005 \text{ in}} = +1$$

From App. 12.A, the area outside $z = +1$ is

$$0.5 - 0.3413 = 0.1587$$

Because these are symmetrical limits, $z_{\text{lower}} = -1$.

$$\text{total fraction defective} = (2)(0.1587)$$
$$= \boxed{0.3174 \quad (0.32)}$$

The answer is (C).

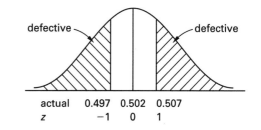

(b) This is a binomial problem.

$$p = p\{\text{defective}\} = 0.3174$$
$$q = 1 - p = 0.6826$$

From Eq. 12.28,

$$p\{x\} = f(x) = \binom{n}{x}\hat{p}^x - \hat{q}^{n-x}$$
$$f(2) = \binom{15}{2}(0.3174)^2(0.6826)^{13}$$
$$= \left(\frac{15!}{13!2!}\right)(0.3174)^2(0.6826)^{13}$$
$$= \boxed{0.0739 \quad (0.074)}$$

The answer is (A).

9. This is a typical hypothesis test of two sample population means. The two populations are the original population the manufacturer used to determine the 1600 hr average life value and the new population the sample was taken from. The mean ($\bar{x} = 1520$ hr) of the sample and its standard deviation ($s = 120$ hr) are known, but the mean and standard deviation of a population of average lifetimes are unknown.

Assume that the average lifetime population mean and the sample mean are identical.

$$\bar{x} = \mu = 1520 \text{ hr}$$

The standard deviation of the average lifetime population is

$$\sigma_{\bar{x}} = \frac{s}{\sqrt{n}} = \frac{120 \text{ hr}}{\sqrt{100}} = 12 \text{ hr}$$

The manufacturer can be reasonably sure that the claim of a 1600 hr average life is justified if the average test life is near 1600 hr. "Reasonably sure" must be evaluated based on acceptable probability of being incorrect. If the manufacturer is willing to be wrong with a 5% probability, then a 95% confidence level is required.

Since the direction of bias is known, a one-tailed test is required. To determine if the mean has shifted downward, test the hypothesis that 1600 hr is within the 95% limit of a distribution with a mean of 1520 hr and a standard deviation of 12 hr. From a standard normal table, 5% of a standard normal distribution is outside of z equals 1.645.

$$\text{confidence limit} = \bar{x} + z\sigma_x$$

Therefore,

$$95\% \text{ confidence limit} = 1520 \text{ hr} + (1.645)(12 \text{ hr})$$
$$= 1540 \text{ hr}$$

The manufacturer can be 95% certain that the average lifetime of the bearings is less than 1600 hr.

If the manufacturer is willing to be wrong with a probability of only 1%, then a 99% confidence limit is required. From the normal table, $z = 2.33$, and the 99% confidence limit is

$$99\% \text{ confidence limit} = 1520 \text{ hr} + (2.33)(12 \text{ hr})$$
$$= 1548 \text{ hr}$$

> The manufacturer can be 99% certain that the average bearing life is less than 1600 hr.

The answer is (D).

10. (a) Plot the data points to determine if the relationship is linear.

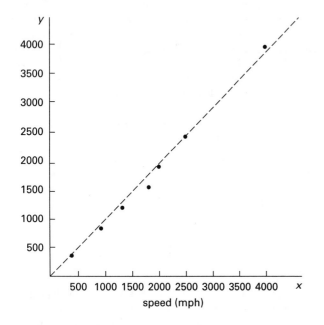

The data appear to be essentially linear. The slope, m, and the y intercept, b, can be determined using linear regression.

The individual terms are

$$n = 7$$
$$\sum x_i = 400 + 800 + 1250 + 1600 + 2000 + 2500$$
$$+ 4000$$
$$= 12{,}550$$
$$\left(\sum x_i\right)^2 = (12{,}550)^2$$
$$= 1.575 \times 10^8$$
$$\bar{x} = \frac{\sum x_i}{n}$$
$$= \frac{12{,}550}{7}$$
$$= 1792.9$$
$$\sum x_i^2 = (400)^2 + (800)^2 + (1250)^2 + (1600)^2$$
$$+ (2000)^2 + (2500)^2 + (4000)^2$$
$$= 3.117 \times 10^7$$

Similarly,

$$\sum y_i = 370 + 780 + 1210 + 1560 + 1980$$
$$+ 2450 + 3950$$
$$= 12{,}300$$

$$\left(\sum y_i\right)^2 = (12{,}300)^2 = 1.513 \times 10^8$$
$$\bar{y} = \frac{\sum y_i}{n} = \frac{12{,}300}{7} = 1757.1$$
$$\sum y_i^2 = (370)^2 + (780)^2 + (1210)^2 + (1560)^2$$
$$+ (1980)^2 + (2450)^2 + (3950)^2$$
$$= 3.017 \times 10^7$$

Also,

$$\sum x_i y_i = (400)(370) + (800)(780) + (1250)(1210)$$
$$+ (1600)(1560) + (2000)(1980)$$
$$+ (2500)(2450) + (4000)(3950)$$
$$= 3.067 \times 10^7$$

The slope is

$$m = \frac{n\sum x_i y_i - \sum x_i \sum y_i}{n\sum x_i^2 - \left(\sum x_i\right)^2}$$

$$= \frac{(7)(3.067 \times 10^7) - (12{,}550)(12{,}300)}{(7)(3.117 \times 10^7) - (12{,}550)^2}$$

$$= 0.994$$

The y-intercept is

$$b = \bar{y} - m\bar{x}$$

$$= 1757.1 - (0.994)(1792.9)$$

$$= -25.0$$

The least squares equation of the line is

$$y = mx + b$$

$$= \boxed{0.994x - 25.0}$$

(b) The correlation coefficient is

$$r = \frac{n\sum x_i y_i - \sum x_i \sum y_i}{\sqrt{\left(n\sum x_i^2 - \left(\sum x_i\right)^2\right)\left(n\sum y_i^2 - \left(\sum y_i\right)^2\right)}}$$

$$= \frac{(7)(3.067 \times 10^7) - (12{,}500)(12{,}300)}{\sqrt{\begin{array}{c}\left((7)(3.117 \times 10^7) - (12{,}500)^2\right) \\ \times \left((7)(3.017 \times 10^7) - (12{,}300)^2\right)\end{array}}}$$

$$\approx \boxed{1.00}$$

11. Plotting the data shows that the relationship is nonlinear.

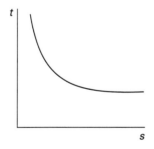

This appears to be an exponential with the form

$$t = ae^{bs}$$

Take the natural log of both sides.

$$\ln t = \ln(ae^{bs})$$

$$= \ln a + \ln(e^{bs})$$

$$= \ln a + bs$$

But, $\ln a$ is just a constant, c.

$$\ln t = c + bs$$

Make the transformation $R = \ln t$.

$$R = c + bs$$

s	R
20	3.76
18	4.95
16	5.95
14	7.00

This is linear.

$$n = 4$$

$$\sum s_i = 20 + 18 + 16 + 14 = 68$$

$$\bar{s} = \frac{\sum s}{n} = \frac{68}{4} = 17$$

$$\sum s_i^2 = (20)^2 + (18)^2 + (16)^2 + (14)^2 = 1176$$

$$\left(\sum s_i\right)^2 = (68)^2 = 4624$$

$$\sum R_i = 3.76 + 4.95 + 5.95 + 7.00 = 21.66$$

$$\bar{R} = \frac{\sum R_i}{n} = \frac{21.66}{4} = 5.415$$

$$\sum R_i^2 = (3.76)^2 + (4.95)^2 + (5.95)^2 + (7.00)^2$$

$$= 123.04$$

$$\left(\sum R_i\right)^2 = (21.66)^2 = 469.16$$

$$\sum s_i R_i = (20)(3.76) + (18)(4.95) + (16)(5.95)$$

$$+ (14)(7.00)$$

$$= 357.5$$

The slope, b, of the transformed line is

$$b = \frac{n\sum s_i R_i - \sum s_i \sum R_i}{n\sum s_i^2 - \left(\sum s_i\right)^2}$$

$$= \frac{(4)(357.5) - (68)(21.66)}{(4)(1176) - (68)^2}$$

$$= -0.536$$

The intercept is

$$c = \bar{R} - b\bar{s}$$
$$= 5.415 - (-0.536)(17)$$
$$= 14.527$$

The transformed equation is

$$R = c + bs$$
$$= 14.527 - 0.536s$$
$$\boxed{\ln t = 14.527 - 0.536s}$$

12. The first step is to graph the data.

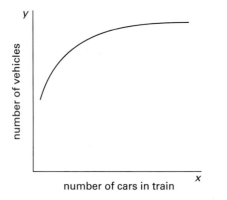

y-axis: number of vehicles
x-axis: number of cars in train

It is assumed that the relationship between the variables has the form $y = a + b\log x$. Therefore, the variable change $z = \log x$ is made, resulting in the following set of data.

z	y
0.301	14.8
0.699	18.0
0.903	20.4
1.079	23.0
1.431	29.9

$$\sum z_i = 4.413$$
$$\sum y_i = 106.1$$
$$\sum z_i^2 = 4.6082$$
$$\sum y_i^2 = 2382.2$$
$$\left(\sum z_i\right)^2 = 19.475$$
$$\left(\sum y_i\right)^2 = 11{,}257.2$$
$$\bar{z} = 0.8826$$
$$\bar{y} = 21.22$$
$$\sum z_i y_i = 103.06$$
$$n = 5$$

The slope is

$$m = \frac{n\sum z_i y_i - \sum z_i \sum y_i}{n\sum z_i^2 - \left(\sum z_i\right)^2}$$
$$= \frac{(5)(103.06) - (4.413)(106.1)}{(5)(4.6082) - 19.475}$$
$$= 13.20$$

The y-intercept is

$$b = \bar{y} - m\bar{z}$$
$$= 21.22 - (13.20)(0.8826)$$
$$= 9.570$$

The resulting equation is

$$y = 9.570 + 13.20z$$

The relationship between x and y is approximately

$$\boxed{y = 9.570 + 13.20\log x}$$

This is not an optimal correlation, as better correlation coefficients can be obtained if other assumptions about the form of the equation are made. For example, $y = 9.1 + 4\sqrt{x}$ has a better correlation coefficient.

13. (a) Plot the data to verify that they are linear.

y-axis: Y − 90
x-axis: T − 200

x $T - 200$	y $Y - 90$
7.1	2.30
10.3	2.58
0.4	1.56
1.1	1.63
3.4	1.83

step 1:

$$\sum x_i = 22.3 \qquad \sum y_i = 9.9$$
$$\sum x_i^2 = 169.43 \qquad \sum y_i^2 = 20.39$$
$$\left(\sum x_i\right)^2 = 497.29 \qquad \left(\sum y_i\right)^2 = 98.01$$
$$\bar{x} = \frac{22.3}{5} = 4.46 \qquad \bar{y} = 1.98$$
$$\sum x_i y_i = 51.54$$

step 2: From Eq. 12.70, the slope is

$$m = \frac{n\sum(x_i y_i) - (\sum x_i)(\sum y_i)}{n\sum x_i^2 - (\sum x_i)^2}$$

$$= \frac{(5)(51.54) - (22.3)(9.9)}{(5)(169.43) - 497.29} = 0.1055$$

step 3: From Eq. 12.71, the y-intercept is

$$b = \bar{y} - m\bar{x}$$

$$= 1.98 - (0.1055)(4.46) = 1.509$$

The equation of the line is

$$y = 0.1055x + 1.509$$

$$Y - 90 = 0.1055(T - 200) + 1.509$$

$$Y = \boxed{0.1055\,T + 70.409}$$

(b) *step 4:* Use Eq. 12.72 to get the correlation coefficient.

$$r = \frac{n\sum(x_i y_i) - (\sum x_i)(\sum y_i)}{\sqrt{\left(n\sum x_i^2 - (\sum x_i)^2\right)\left(n\sum y_i^2 - (\sum y_i)^2\right)}}$$

$$= \frac{(5)(51.54) - (22.3)(9.9)}{\sqrt{((5)(169.43) - 497.29)((5)(20.39) - 98.01)}}$$

$$= \boxed{0.995}$$

14. Plot the data to see if they are linear.

This looks like it could be of the form

$$y = a + b\sqrt{x}$$

However, when x is negative (as in the first point), the function is imaginary. Try shifting the curve to the right, replacing x with $x + 1$.

$$y = a + bz$$

$$z = \sqrt{x + 1}$$

z	y
0	0
1	1
1.414	1.4
1.732	1.7
2	2
2.236	2.2
2.45	2.4
2.65	2.6
2.83	2.8
3	3

Because $y \approx z$, the relationship is

$$\boxed{y = \sqrt{x + 1}}$$

In this problem, the answer was found with some good guesses. Usually, regression would be necessary.

15. (a)

R (Ω)	f	fR (Ω)	fR^2 (Ω^2)
0.200	1	0.200	0.0400
0.210	3	0.630	0.1323
0.220	5	1.100	0.2420
0.230	10	2.300	0.5290
0.240	17	4.080	0.9792
0.250	40	10.000	2.5000
0.260	13	3.380	0.8788
0.270	6	1.620	0.4374
0.280	3	0.840	0.2352
0.290	2	0.580	0.1682
	100	24.730	6.1421

$$\bar{R} = \frac{\sum fR}{\sum f} = \frac{24.730\ \Omega}{100} = \boxed{0.2473\ \Omega \quad (0.247\ \Omega)}$$

The answer is (B).

(b) The sample standard deviation is given by Eq. 12.61.

$$s = \sqrt{\dfrac{\sum fR^2 - \dfrac{\left(\sum fR\right)^2}{n}}{n-1}}$$

$$= \sqrt{\dfrac{6.1421 \ \Omega - \dfrac{(24.73 \ \Omega)^2}{100}}{99}}$$

$$= \boxed{0.0163 \ \Omega \quad (0.016 \ \Omega)}$$

The answer is (C).

(c) 50% of the observations are below the median resistance, R_{median}. 36 observations are below 0.240, so 14 more are needed.

$$R_{\text{median}} = R_{36} + f_{\text{interval}} \, \Delta R_{\text{interval}}$$

$$= 0.240 \ \Omega + \left(\frac{14}{40}\right)(0.250 \ \Omega - 0.240 \ \Omega)$$

$$= \boxed{0.2435 \ \Omega \quad (0.244 \ \Omega)}$$

The answer is (B).

(d)
$$s^2 = (0.0163 \ \Omega)^2$$

$$= \boxed{0.0002656 \ \Omega^2 \quad (0.00027 \ \Omega^2)}$$

The answer is (A).

13 Computer Mathematics

PRACTICE PROBLEMS

1. What is the base-10 value of the base-2 number 1010?

   (A) 3_{10}

   (B) 7_{10}

   (C) 10_{10}

   (D) 20_{10}

2. What is the value of 144_{10} in base-2?

   (A) $1\,001\,000_2$

   (B) $1\,010\,000_2$

   (C) $10\,010\,000_2$

   (D) $10\,010\,001_2$

3. What is the value of 144_{10} in base-8, or octal, numbering?

   (A) 22_8

   (B) 202_8

   (C) 220_8

   (D) 221_8

4. What is the base-10 value of the hexadecimal number $(3C1)_{16}$?

   (A) 192_{10}

   (B) 768_{10}

   (C) 961_{10}

   (D) $15\,360_{10}$

5. What is the one's complement value of $101\,011_2$?

   (A) $10\,010_2$

   (B) $10\,100_2$

   (C) $10\,111_2$

   (D) $101\,011_2$

SOLUTIONS

1. For a positional numbering system

$$(a_n a_{n-1} \cdots a_2 a_1 a_0) = a_n b^n + a_{n-1} b_{n-1} + \cdots$$
$$+ a_2 b^2 + a_1 b + a_0$$

Therefore, for $b = 2$

$$(1010)_2 = (1)(2^3) + (0)(2^2) + (1)(2) + 0$$
$$= 8 + 0 + 2 + 0$$
$$= \boxed{10_{10}}$$

The answer is (C).

2. Using the remainder method

$$
\begin{array}{ll}
144/2 = 72 & \text{remainder } 0 \\
72/2 = 36 & \text{remainder } 0 \\
36/2 = 18 & \text{remainder } 0 \\
18/2 = 9 & \text{remainder } 0 \\
9/2 = 4 & \text{remainder } 1 \\
4/2 = 2 & \text{remainder } 0 \\
2/2 = 1 & \text{remainder } 0 \\
1/2 = 0 & \text{remainder } 1 \\
\end{array}
$$

Reading up the remainder values from the last division remainder of 1 to the first remainder gives a base-2 value of $10\,010\,000_2$. The binary representation of 144_{10} is therefore $\boxed{10\,010\,000_2.}$

The answer is (C).

3. Using the remainder method gives

$$
\begin{array}{ll}
144/8 = 18 & \text{remainder } 0 \\
18/8 = 2 & \text{remainder } 2 \\
2/8 = 0 & \text{remainder } 2 \\
\end{array}
$$

Reading up the remainder values from the last division remainder of 1 to the first remainder gives a base-8 value of 220_8. The octal representation of 144_{10} is $\boxed{220_8.}$

The answer is (C).

4. For a position numbering system

$$(a_n a_{n-1} \cdots a_2 a_1 a_0) = a_n b^n + a_{n-1} b_{n-1} + \cdots$$
$$+ a_2 b^2 + a_1 b + a_0$$

Therefore, in base-10 representation, for $b = 16$,

$$(3C1)_{16} = (3)(16)^2 + (12)(16) + 1$$
$$= 768 + 192 + 1$$
$$= \boxed{961_{10}}$$

The answer is (C).

5. The one's complement is

$$N_1^x = N_2^x - 1$$

The one's complement is found by subtracting 1 from the two's complement value. Or, the one's complement can be found by reversing all the digits in the base-2 number. That is, all 0's become 1's and vice versa. Therefore, for $101\,011_2$. the one's complement is $010\,100_2$ or, dropping the leading 0, $\boxed{10\,100_2.}$

The answer is (B).

14 Numerical Analysis

PRACTICE PROBLEMS

1. A function is given as $y = 3x^{0.93} + 4.2$. What is the percent error if the value of y at $x = 2.7$ is found by using straight-line interpolation between $x = 2$ and $x = 3$?

(A) 0.060%

(B) 0.18%

(C) 2.5%

(D) 5.4%

2. Given the following data points, find y by straight-line interpolation for $x = 2.75$.

x	y
1	4
2	6
3	2
4	−14

(A) 2.1

(B) 2.4

(C) 2.7

(D) 3.0

3. Using the bisection method, find all the roots of $f(x) = 0$, to the nearest 0.000005.

$$f(x) = x^3 + 2x^2 + 8x - 2$$

4. What is most nearly the root of the following equation?

$$f(x) = x^3 - 4x - 5$$

Use Newton's method with $x_0 = 2$.

(A) 2.47

(B) 2.60

(C) 2.63

(D) 16.7

5. What is the primary DISADVANTAGE of performing a straight-line interpolation—as opposed to performing a polynomial interpolation—using either the method of Lagrange or Newton?

(A) Calculations are complicated.

(B) Curvature is ignored.

(C) Iterations are required.

(D) Process is slow.

SOLUTIONS

1. The actual value at $x = 2.7$ is

$$y(x) = 3x^{0.93} + 4.2$$
$$y(2.7) = (3)(2.7)^{0.93} + 4.2$$
$$= 11.756$$

At $x = 3$,

$$y(3) = (3)(3)^{0.93} + 4.2$$
$$= 12.534$$

At $x = 2$,

$$y(2) = (3)(2)^{0.93} + 4.2$$
$$= 9.916$$

Use straight-line interpolation.

$$\frac{x_2 - x}{x_2 - x_1} = \frac{y_2 - y}{y_2 - y_1}$$
$$\frac{3 - 2.7}{3 - 2} = \frac{12.534 - y}{12.534 - 9.916}$$
$$y = 11.749$$

The relative error is given by

$$\frac{\text{actual value} - \text{predicted value}}{\text{actual value}} = \frac{11.756 - 11.749}{11.756}$$
$$= \boxed{0.000595 \quad (0.060\%)}$$

The answer is (A).

2. Let $x_1 = 2$; from the table of data points, $y_1 = 6$. Let $x_2 = 3$; from the table of data points, $y_2 = 2$.

Let $x = 2.75$. By straight-line interpolation,

$$\frac{x_2 - x}{x_2 - x_1} = \frac{y_2 - y}{y_2 - y_1}$$
$$\frac{3 - 2.75}{3 - 2} = \frac{2 - y}{2 - 6}$$
$$\boxed{y = 3}$$

The answer is (D).

3. $f(x) = x^3 + 2x^2 + 8x - 2$

Try to find an interval in which there is a root.

x	$f(x)$
0	-2
1	9

A root exists in the interval $[0, 1]$.

Try $x = \left(\frac{1}{2}\right)(0 + 1) = 0.5$.

$$f(0.5) = (0.5)^3 + (2)(0.5)^2 + (8)(0.5) - 2 = 2.625$$

A root exists in $[0, 0.5]$.

Try $x = 0.25$.

$$f(0.25) = (0.25)^3 + (2)(0.25)^2 + (8)(0.25) - 2 = 0.1406$$

A root exists in $[0, 0.25]$.

Try $x = 0.125$.

$$f(0.125) = (0.125)^3 + (2)(0.125)^2 + (8)(0.125) - 2$$
$$= -0.967$$

A root exists in $[0.125, 0.25]$.

Try $x = \left(\frac{1}{2}\right)(0.125 + 0.25) = 0.1875$.

Continuing,

$$f(0.1875) = -0.42 \quad [0.1875, 0.25]$$
$$f(0.21875) = -0.144 \quad [0.21875, 0.25]$$
$$f(0.234375) = -0.002 \quad [\text{This is close enough.}]$$

One root is $x_1 \approx \boxed{0.234375.}$

Try to find the other two roots. Use long division to factor the polynomial.

$$
\require{enclose}
\begin{array}{r}
x^2 + 2.234375x + 8.52368 \\[-3pt]
x - 0.234375 \enclose{longdiv}{\, x^3 + \quad 2x^2 + \quad\quad 8x - 2} \\[-3pt]
\underline{-(x^3 - 0.234375x^2)} \\[-3pt]
2.234375x^2 + \quad\quad 8x \\[-3pt]
\underline{-(2.234375x^2 - 0.52368x)} \\[-3pt]
8.52368x - 2 \\[-3pt]
\underline{-(8.52368x - 1.9977)} \\[-3pt]
\approx 0
\end{array}
$$

Use the quadratic equation to find the roots of $x^2 + 2.234375x + 8.52368$.

$$x_2, x_3 = \frac{-2.234375 \pm \sqrt{(2.234375)^2 - (4)(1)(8.52368)}}{(2)(1)}$$

$$= \boxed{-1.117189 \pm j2.697327} \quad \text{[both imaginary]}$$

4. Newton's method for finding a root is

$$x_{n+1} = g(x_n) = x_n - \frac{f(x_n)}{f'(x_n)}$$

The function and its first derivative are

$$f(x) = x^3 - 4x - 5$$
$$f'(x) = 3x^2 - 4$$

The first iteration, with $n = 0$, gives

$$x_0 = 2$$
$$f(x_0) = f(2) = 2^3 - (4)(2) - 5 = -5$$
$$f'(x_0) = f'(2) = (3)(2)^2 - 4 = 8$$
$$x_1 = x_0 - \frac{f(x_0)}{f'(x_0)}$$
$$= 2 - \left(\frac{-5}{8}\right)$$
$$= 2.625$$

The second iteration, with $n = 1$, gives

$$x_1 = 2.625$$
$$f(x_1) = f(2.625) = 2.625^3 - (4)(2.625) - 5 = 2.588$$
$$f'(x_1) = f'(2.625) = (3)(2.625)^2 - 4 = 16.672$$
$$x_2 = x_1 - \frac{f(x_1)}{f'(x_1)}$$
$$= 2.625 - \left(\frac{2.588}{16.672}\right)$$
$$= \boxed{2.470 \quad (2.47)}$$

The iterations may be continued to the desired accuracy.

The answer is (A).

5. The straight-line interpolation technique ignores the effects of curvature. Lagrangian and Newtonian polynomial methods of interpolation do not.

The answer is (B).

15 Advanced Engineering Mathematics

PRACTICE PROBLEMS

1. What is the correct description of the following differential equation?

$$2y''(t) + 3y'(t) + 1 = f(t)$$

(A) linear, first order, homogeneous

(B) nonlinear, first order, homogeneous

(C) linear, second order, nonhomogeneous

(D) nonlinear, second order, nonhomogeneous

2. What is the characteristic equation for the following differential equation?

$$3y''(t) + 5y'(t) + 6y(t) = 24t$$

(A) $3\lambda^2 + 5\lambda + 6 = 0$

(B) $3\lambda^2 + 5\lambda + 6 = 24$

(C) $3\lambda^2 + 5\lambda = 18$

(D) $3\lambda^2 + 5\lambda = 24$

3. Which of the following is primarily used to analyze periodic signals when $s = j\omega$?

(A) Fourier series

(B) Fourier transforms

(C) Laplace transforms

(D) z-transforms

4. What is the Laplace transform of the following?

$$x(t - t_0)$$

(A) $X(s)$

(B) $e^{-st}X(s)$

(C) e^{-st_0}

(D) $e^{-st_0}X(s)$

5. What type of function is shown?

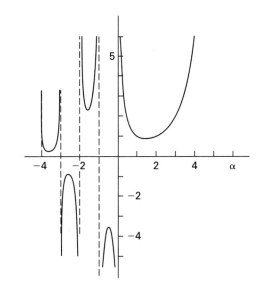

(A) delta function

(B) error function

(C) gamma function

(D) unit finite impulse function

Mathematics

SOLUTIONS

1. All the coefficients of the terms are constants, therefore, the equation is linear. Because there is a second derivative, the equation is of second order. Because the forcing function, $f(t)$, exists, the implication is that the value of $f(t)$ is non-zero. Therefore, the equation is nonhomogeneous.

The answer is (C).

2. The characteristic equation is

$$\boxed{3\lambda^2 + 5\lambda + 6 = 0}$$

The characteristic equation is used to solve the homogeneous differential equation.

The answer is (A).

3. Fourier series and Laplace transforms are used to analyze periodic signals. The Laplace transform is more general in nature.

$$\mathcal{L}\big(f(t)\big)\Big|_{s=j\omega} = F(s)\Big|_{s=j\omega} = \mathcal{F}\big(f(t)\big)$$

$$\boxed{\text{Therefore, the Fourier series is used.}}$$

The Fourier series is equivalent to the Laplace transform when $s = j\omega$, that is, when a transient condition is not present. When $s = \sigma + j\omega$, a transient condition is present, represented by the σ term, and the Fourier series is not equivalent to the Laplace transform.

The answer is (A).

4. The Laplace transform of a time-shifted function is

$$\boxed{e^{-st_0} X(s)}$$

The answer is (D).

5. The illustration shows the factorial function, more commonly called the gamma function.

$$\Gamma(\alpha) = \int_0^\infty e^{-t} t^{(\alpha-1)} \, dt \quad [\alpha > 0]$$

The answer is (C).

16 Electromagnetic Theory

PRACTICE PROBLEMS

1. What is the magnitude of the electric force between a proton and electron 0.046 nm apart in free space?

(A) 9.651×10^{-19} N

(B) -5.018×10^{-18} N

(C) 5.018×10^{-18} N

(D) 1.090×10^{-7} N

2. What is the magnitude of the electric field strength for a proton in free space at 0.046 nm?

(A) 3.047×10^{-30} N/C

(B) 1.440×10^{-9} N/C

(C) 31.30 N/C

(D) 6.805×10^{11} N/C

3. What is the magnitude of the electric field intensity for a parallel plate capacitor with a potential of 12 V applied and a distance of 0.3 cm between plates?

(A) 0.12 V/m

(B) 0.40 V/m

(C) 40 V/m

(D) 4000 V/m

4. During design discussion, quartz is picked as the dielectric for certain capacitors. What is the expected permittivity?

(A) 1.8×10^{-12} C^2/N·m^2

(B) 8.8×10^{-12} C^2/N·m^2

(C) 4.4×10^{-11} C^2/N·m^2

(D) 8.8×10^{-8} C^2/N· m^2

5. What is the magnitude of the electric flux density 0.5 m from a wire with a uniform line charge of 30×10^{-5} C/m?

(A) 9.5×10^{-5} C/m^2

(B) 1.9×10^{-4} C/m^2

(C) 6.0×10^{-4} C/m^2

(D) 1.7×10^{-3} C/m^2

SOLUTIONS

1. The magnitude of the electric force is given by Coulomb's law. From Eq. 16.3,

$$F = \frac{|Q||Q|}{4\pi\epsilon_0 r^2}$$

$$= \frac{(1.6022 \times 10^{-19}\ \text{C})(1.6022 \times 10^{-19}\ \text{C})}{4\pi\left(8.854 \times 10^{-12}\ \dfrac{\text{C}^2}{\text{N·m}^2}\right)(0.046 \times 10^{-9}\ \text{m})^2}$$

$$= \boxed{1.090 \times 10^{-7}\ \text{N}}$$

0.046 nm is the approximate distance between the proton and electron in the hydrogen atom.

The answer is (D).

2. From Eq. 16.6, the magnitude of the electric field strength, E, is

$$E = \frac{Q}{4\pi\epsilon_0 r^2}$$

$$= \frac{1.6022 \times 10^{-19}\ \text{C}}{4\pi\left(8.854 \times 10^{-12}\ \dfrac{\text{C}^2}{\text{N·m}^2}\right)(0.046 \times 10^{-9}\ \text{m})^2}$$

$$= \boxed{6.805 \times 10^{11}\ \text{N/C}}$$

The answer is (D).

3. From Table 16.6, the electric field intensity for a parallel plate capacitor (ignoring fringing effects) is

$$E = \frac{V}{r}$$

$$= \frac{12\ \text{V}}{(0.3\ \text{cm})\left(\dfrac{1\ \text{m}}{100\ \text{cm}}\right)}$$

$$= \boxed{4000\ \text{V/m}}$$

The answer is (D).

4. From Table 16.7,

$$\epsilon_r = 5$$

From Eq. 16.11, the permittivity is

$$\epsilon = \epsilon_0 \epsilon_r$$

$$= \left(8.854 \times 10^{-12} \ \frac{C^2}{N \cdot m^2}\right)(5)$$

$$= \boxed{4.4 \times 10^{-11} \ C^2/N \cdot m^2}$$

The answer is (C).

5. From Eq. 16.21, the electric flux density for a uniform line charge is

$$D = \frac{\rho_l}{2\pi r}$$

$$= \frac{30 \times 10^{-5} \ \dfrac{C}{m}}{2\pi(0.5 \ m)}$$

$$= \boxed{9.5 \times 10^{-5} \ C/m^2}$$

30×10^{-5} C/m correlates approximately with 30 A using an electron speed of 10^5 m/s, which is the thermal limit for electrons in silicon at room temperature.

The answer is (A).

17 Electronic Theory

PRACTICE PROBLEMS

1. Assuming no kinetic energy will be imparted to the electron, what frequency of incoming radiation is required to free an electron from the surface of silicon that has a work function of 4.8 V?

(A) 5.6×10^4 Hz

(B) 1.2×10^{15} Hz

(C) 3.5×10^{23} Hz

(D) 7.2×10^{33} Hz

2. Using the properties of germanium at room temperature, what is the conductivity of the holes?

(A) 7.6×10^{-7} S/m

(B) 7.6×10^{-2} S/m

(C) 4.8×10^{14} S/m

(D) 9.5×10^{14} S/m

3. Silicon and germanium are widely used in semiconductors. How many valence electrons exist for a given atom of either?

(A) 2

(B) 4

(C) 6

(D) 10

4. In an n-type silicon crystal, the number of electrons is

(A) equal to the number of holes

(B) greater than the number of holes

(C) less than the number of holes

(D) equal to the square of the number of holes

5. The n_i of gallium arsenide is 1.1×10^7 cm^{-3}. If the hole concentration is 1.5×10^2 cm^{-3}, what is the electron concentration?

(A) 1.5×10^2 cm^{-3}

(B) 7.3×10^4 cm^{-3}

(C) 1.1×10^7 cm^{-3}

(D) 8.1×10^{11} cm^{-3}

SOLUTIONS

1. From Eq. 17.6,

$$h\nu = \varphi + \tfrac{1}{2}mv^2$$

From App. 17.A, the work function for silicon is 4.8 V. Substituting into the equation gives

$$\frac{\varphi}{h} = \frac{(4.8 \text{ V})\left(1 \frac{\text{J}}{\text{C·V}}\right)\left(1.6022 \times 10^{-19} \frac{\text{C}}{\text{electron}}\right)}{6.6256 \times 10^{-34} \text{ J·s}}$$

$$= \boxed{1.2 \times 10^{15} \text{ Hz}}$$

The answer is (B).

2. From Eq. 17.8,

$$\sigma_h = \rho_h \mu_h$$

From App. 17.A,

$$n_i = 2.5 \times 10^{12} \text{ cm}^{-3} = n_{\text{holes}}$$

$$q_h = 1.6022 \times 10^{-19} \text{ C}$$

Because in an intrinsic semiconductor there are equal numbers of electrons and holes,

$$\rho_h = n_i q_h$$
$$= (2.5 \times 10^{12} \text{ cm}^{-3})(1.6022 \times 10^{-19} \text{ C})$$
$$= 4.0 \times 10^{-7} \text{ C/cm}^3$$

From App. 17.A,

$$\mu_h = 1900 \text{ cm}^2/\text{V·s}$$

$$\sigma = \rho_h \mu_h$$
$$= \left(4.0 \times 10^{-7} \frac{\text{C}}{\text{cm}^3}\right)\left(1900 \frac{\text{cm}^2}{\text{V·s}}\right)\left(100 \frac{\text{cm}}{\text{m}}\right)$$
$$= \boxed{7.6 \times 10^{-2} \text{ S/m}}$$

The answer is (B).

3. Silicon has an electronic structure of

$$1s^2 2s^2 2p^6 3s^2 3p^2$$

Germanium has an electronic structure of

$$1s^2 2s^2 2p^6 3s^2 3p^6 3d^{10} 4s^2 4p^2$$

Both have four valence electrons in the outer shell.

The answer is (B).

4. An n-type crystal is doped with donor impurities. This increases the charge carrier density of electrons, and thereby reduces the number of holes. By Eq. 17.13,

$$np = n_i^2 = \text{constant}$$

The answer is (B).

5. Using the law of mass action, Eq. 17.13,

$$np = n_i^2$$

$$n = \frac{n_i^2}{p} = \frac{(1.1 \times 10^7 \text{ cm}^{-3})^2}{1.5 \times 10^2 \text{ cm}^{-3}}$$

$$= \boxed{8.1 \times 10^{11} \text{ cm}^{-3}}$$

The answer is (D).

18 Communication Theory

PRACTICE PROBLEMS

1. A 60 Hz signal is to be transformed into a digital signal. Determine the ideal sampling rate.

2. Determine the Nyquist interval for a 400 Hz signal.

3. A certain signal has an equivalent rectangular bandwidth of 6 MHz. Determine the response time of the system.

4. A 32-bit computer system has equal probabilities for each of its information elements, x_i, or bits. Determine this probability.

5. Describe entropy as used in information theory.

6. The signal power is doubled at a given transmitter. What is the decibel increase?

    (A) 0.3 dB

    (B) 2 dB

    (C) 3 dB

    (D) 20 dB

7. A certain communication signal uses a bandwidth, BW, of 2 kHz at 101 MHz. What is the classification of this signal?

    (A) narrowband, VLF

    (B) voice-grade, HF

    (C) voice-grade, VHF

    (D) wideband, UHF

8. A certain communication system uses the left-most bit as an even parity bit. Which of the following messages is in error?

    (A) 11101

    (B) 00101

    (C) 10001

    (D) 01011

9. What is the maximum thermal noise voltage generated in a 5 kΩ resistor operating in a 6 MHz bandwidth and designed for a temperature service range of $-10°C$ to $60°C$?

    (A) 9.71×10^{-6} V

    (B) 1.14×10^{-5} V

    (C) 2.35×10^{-5} V

    (D) 2.46×10^{-5} V

10. What is the shot noise current for a microphone with a 16 kHz bandwidth drawing 500 mA?

    (A) 2.60×10^{-12} A

    (B) 2.53×10^{-8} A

    (C) 2.56×10^{-8} A

    (D) 5.06×10^{-8} A

SOLUTIONS

1. The ideal sampling rate for a signal of frequency f_I is given by the Nyquist rate, f_S.

From Eq. 18.1,

$$f_S = 2f_I$$
$$= (2)(60 \text{ Hz})$$
$$= \boxed{120 \text{ Hz}}$$

In practice, the rate would have to be greater than this to avoid a zero output, which would occur if the initial sampling occurred at a value of zero for the analog signal.

2. From Eq. 18.2, the Nyquist interval, T_S, is

$$T_S = \frac{1}{2f_I}$$

The information signal, or message signal, frequency is given as 400 Hz. Therefore,

$$T_S = \frac{1}{(2)(400 \text{ Hz})}$$
$$= \boxed{0.00125 \text{ s} \quad (1.25 \text{ ms})}$$

3. From Eq. 18.4, the bandwidth, $\Delta\Omega$, and duration or response time, Δt, are related by the following uncertainty relationship.

$$\Delta t \Delta\Omega = 2\pi$$

Rearranging and substituting,

$$\Delta t = \frac{2\pi}{\Delta\Omega} = \frac{2\pi}{6 \times 10^6 \text{ Hz}}$$
$$= \boxed{1.047 \times 10^{-6} \text{ s}}$$

4. The information content of a message is given as 32 bits. The information content is related to the probability as follows.

From Eq. 18.6,

$$I(x_i) = \log_2 \frac{1}{p(x_i)}$$

Rearranging and substituting,

$$p(x_i) = \frac{1}{\text{antilog}_2 I(x_i)}$$
$$= \frac{1}{\text{antilog}_2 (32)}$$
$$= \boxed{\frac{1}{4.295 \times 10^9}}$$

5. Entropy is the mean or average value of a message. It can be considered a measure of the disorder of a system, or of uncertainty in a message. For a given message of alphabet size A, the entropy is determined by Eq. 18.7.

$$H(X) = \sum_{i=1}^{A} p(x_i)I(x_i)$$

Here $p(x_i)$ is the probability of the piece of information x_i. Entropy is normally given in base 2, in units of bits. In base e, the unit is the nat.

6. The decibel is defined by the following ratio of power, P_{dB}.

$$P_{dB} = 10 \log_{10} \frac{P_2}{P_1}$$

Here, P_1 is a reference power and P_2 is the signal power. Because the signal power is doubled, P_2/P_1 equals 2.

$$P_{dB} = 10 \log_{10} 2$$
$$= \boxed{3 \text{ dB}}$$

The answer is (C).

7. Classification can mean a number of things. Given the possible choices, this signal is a voice-grade channel (BW is 300 Hz to 4 kHz) that is in the VHF range (30 MHz to 300 MHz).

The answer is (C).

8. The message is contained in the bits to the right of the first bit in the signal, the parity bit. The parity represents an even or odd total of ones. The following table is helpful.

option	message	parity	signal
(A)	1101	1	11101
(B)	0101	0	00101
(C)	0001	1	10001
(D)	1011	1	11011

By comparison, message (D) is in error.

The answer is (D).

9. From Eq. 18.10, the mean squared thermal noise voltage is

$$\overline{v_n^2} = 4kTR(\text{BW})$$
$$v_n = \sqrt{4kTR(\text{BW})}$$

Here, k is Boltzmann's constant, T is the absolute temperature, R is resistance, and BW is the bandwidth. The maximum voltage occurs at the maximum temperature (60°C or 333K), so

$$v = \sqrt{\begin{array}{l}(4)\left(1.38 \times 10^{-23} \, \dfrac{\text{J}}{\text{K}}\right)(333\text{K}) \\ \times (5 \times 10^3 \ \Omega)(6 \times 10^6 \ \text{Hz})\end{array}}$$
$$= \boxed{2.35 \times 10^{-5} \ \text{V}}$$

The answer is (C).

10. From Eq. 18.13, the mean squared shot noise current is given in terms of the change per carrier, q.

$$\overline{i_n^2} = 2qI(\text{BW})$$
$$i_n = \sqrt{2qI(\text{BW})}$$
$$= \sqrt{\begin{array}{l}(2)(1.6022 \times 10^{-19} \ \text{C}) \\ \times (500 \times 10^{-3} \ \text{A})(16 \times 10^3 \ \text{Hz})\end{array}}$$
$$= \boxed{5.06 \times 10^{-8} \ \text{A}}$$

The answer is (D).

19 Acoustic and Transducer Theory

PRACTICE PROBLEMS

1. Explain two differences between sound waves and electromagnetic waves.

2. Define the term "transducer."

3. What is the sound pressure level in air for a signal at the pain sensation threshold of 1 W/m^2?

(A) −120 dB

(B) 3.0 dB

(C) 60 dB

(D) 120 dB

4. Explain the difference between photovoltaic and photoconductive transduction.

5. The reference pressure level, p_0, in air of 20 μPa correlates with the threshold audible frequency of 2000 Hz. What is the wavelength of this sound wave?

(A) 0.17 m

(B) 1.0 m

(C) 6.1 m

(D) 20 m

SOLUTIONS

1. Sound is a longitudinal or compression wave. Electromagnetic waves are transverse. That is, the electric and magnetic fields are at right angles to the direction of propagation. Also, sound waves require a medium in which to travel, while electromagnetic waves do not.

2. A transducer is a communication interface that provides a usable output in response to a specified measurand (i.e., a physical quantity, property, or condition).

3. From Ex. 19.2, the pressure level corresponding to 1 W/m^2 is 29.3 Pa. The reference pressure (corresponding to 0 dB) is defined as 20 μPa. Substituting from Eq. 19.8 gives

$$L_p = 20 \log \frac{p}{p_0}$$
$$= 20 \log \frac{29.3 \text{ Pa}}{20 \times 10^{-6} \text{ Pa}}$$
$$= \boxed{123 \text{ dB} \quad (120 \text{ dB})}$$

The "pain" threshold is nominally 120 dB in most references.

The answer is (D).

4. The primary difference between photovoltaic and photoconductive transduction is that a photovoltaic device uses dissimilar metals and a photoconductive device uses a semiconductor *pn* junction.

5. From Table 19.1, the speed of sound in air is approximately 330 m/s. Substituting this and the given frequency into Eq. 19.1 gives

$$\text{v}_s = f\lambda$$
$$\lambda = \frac{\text{v}_s}{f} = \frac{330 \frac{\text{m}}{\text{s}}}{2000 \text{ Hz}}$$
$$= \boxed{0.165 \text{ m} \quad (0.17 \text{ m})}$$

The answer is (A).

20 Electrostatics

PRACTICE PROBLEMS

1. Determine the magnitude of the electric field necessary to place a 1 N force on an electron.

2. A point charge of 4.8×10^{-19} C is located at a point $(3, 1, 0)$. What is the electric field strength at $(3, 0, 1)$?

(A) 1.5×10^{-9} V/m $\left(\dfrac{-\mathbf{y}}{\sqrt{2}} + \dfrac{\mathbf{z}}{\sqrt{2}}\right)$

(B) 2.2×10^{-9} V/m $\left(\dfrac{-\mathbf{y}}{\sqrt{2}} + \dfrac{\mathbf{z}}{\sqrt{2}}\right)$

(C) 3.1×10^{-9} V/m $(-\mathbf{y} + \mathbf{z})$

(D) 4.3×10^{-9} V/m $(-\mathbf{y} + \mathbf{z})$

3. Water molecules in vapor form have a fractional dipole charge of approximately 5.26×10^{-20} C. If the dipole bond length is 1.17 Å, what is the magnitude of the dipole moment?

(A) 6.2×10^{-32} C·m

(B) 6.2×10^{-30} C·m

(C) 5.3×10^{-20} C·m

(D) 6.1×10^{-20} C·m

4. The potential of a point charge, in spherical coordinates and relative to infinity, is

$$V = \frac{Q}{4\pi\epsilon_0 r}$$

Determine the expression for the electric field strength, **E**.

5. The electric field in an n-type material outside the space charge region is given by

$$\mathbf{E} = 2\mathbf{x} + 2\mathbf{y} + 3\mathbf{z}$$

Determine the charge density, ρ, and explain the result.

SOLUTIONS

1. The magnitude of the force is

$$F = QE$$

Rearranging gives

$$E = \frac{F}{Q}$$

$$= \frac{1 \text{ N}}{1.602 \times 10^{-19} \text{ C}}$$

$$= 6.2 \times 10^{18} \frac{\text{N}}{\text{C}}$$

$$= \boxed{6.2 \times 10^{18} \text{ V/m}}$$

For determining the magnitude of the force and not the direction, the minus sign is not required on the charge.

2.

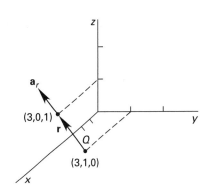

The electric field is

$$\mathbf{E} = \left(\frac{Q}{4\pi\epsilon r^2}\right)\mathbf{a}_r$$

First, determine the value of the unit vector, $\mathbf{a}_r$.

$$a_r = \sqrt{(3-3)^2 + (0-1)^2 + (1-0)^2} = \sqrt{2}$$

$$\mathbf{r} = 0\mathbf{x} - 1\mathbf{y} + 1\mathbf{z}$$

Therefore,

$$\mathbf{a}_r = \frac{\mathbf{r}}{a_r} = \frac{1}{\sqrt{2}}(-\mathbf{y} + \mathbf{z})$$

Substituting this and the given information (free space is assumed) gives

$$\mathbf{E} = \left(\frac{Q}{4\pi\epsilon r^2}\right)\mathbf{a}_r$$

$$= \left(\frac{4.8 \times 10^{-19} \text{ C}}{4\pi\left(8.854 \times 10^{-12} \frac{\text{C}^2}{\text{N·m}^2}\right)(\sqrt{2} \text{ m})^2}\right)$$

$$\times \left(\left(\frac{1}{\sqrt{2}}\right)(-\mathbf{y} + \mathbf{z})\right)$$

$$= \boxed{2.16 \times 10^{-9} \text{ N/C}\left(\frac{-\mathbf{y}}{\sqrt{2}} + \frac{\mathbf{z}}{\sqrt{2}}\right)}$$

The answer is (B).

3. The dipole moment vector, given in terms of a single positive charge, is

$$\mathbf{p} = q\mathbf{d}$$

It points from the negative to positive charge. The separation of charges is d.

An angstrom, Å, equals 10^{-10} m. Substituting gives

$$p = qd = (5.26 \times 10^{-20} \text{ C})(1.17 \times 10^{-10} \text{ m})$$

$$= \boxed{6.15 \times 10^{-30} \text{ C·m} \quad (6.2 \times 10^{-30} \text{ C·m})}$$

The answer is (B).

4. The electric field strength is

$$\mathbf{E} = -\nabla V$$

$$= -\nabla\left(\frac{Q}{4\pi\epsilon_0 r}\right)$$

$$= \left(-\frac{Q}{4\pi\epsilon_0}\right)\left(\nabla\frac{1}{r}\right)$$

In spherical coordinates, the del operator can be represented as

$$\nabla = \left(\frac{\partial}{\partial r}\right)\mathbf{r} + \left(\frac{1}{r}\right)\left(\frac{\partial}{\partial\theta}\right)\boldsymbol{\theta} + \left(\frac{1}{r\sin\theta}\right)\left(\frac{\partial}{\partial\phi}\right)\boldsymbol{\phi}$$

For a point charge, the potential and the electric field are constant with respect to θ and ϕ. Thus,

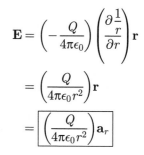

$$\mathbf{E} = \left(-\frac{Q}{4\pi\epsilon_0}\right)\left(\frac{\partial\frac{1}{r}}{\partial r}\right)\mathbf{r}$$

$$= \left(\frac{Q}{4\pi\epsilon_0 r^2}\right)\mathbf{r}$$

$$= \boxed{\left(\frac{Q}{4\pi\epsilon_0 r^2}\right)\mathbf{a}_r}$$

The last equation changes the symbol from $\mathbf{r}$ to $\mathbf{a}_r$ only. Many variations exist.

5. The charge density, ρ, can be determined from Gauss' law.

$$\text{div } \mathbf{E} = \nabla\cdot\mathbf{E} = \frac{\rho}{\epsilon}$$

$$\nabla\cdot\mathbf{E} = \nabla\cdot(2\mathbf{x} + 2\mathbf{y} + 3\mathbf{z})$$

$$= \left(\frac{\partial 2}{\partial x}\right)\mathbf{x} + \left(\frac{\partial 2}{\partial y}\right)\mathbf{y} + \left(\frac{\partial 3}{\partial z}\right)\mathbf{z}$$

$$= 0$$

When the electric field is constant, the divergence is zero. This indicates there is no net charge in the region, as is the case for the pn junction not under the influence of an external electric field.

21 Electrostatic Fields

PRACTICE PROBLEMS

1. Determine the electric flux density magnitude if the polarization is 2.0×10^{-6} C/m^2 and the susceptibility is 2.2.

(A) 1.8×10^{-17} C/m^2

(B) 1.8×10^{-6} C/m^2

(C) 2.9×10^{-6} C/m^2

(D) 4.0×10^{-6} C/m^2

2. Given $\mathbf{D} = (2 \times 10^{-6})\mathbf{a}_r$ C/m^2 and $\epsilon_r = 3.0$, what is the electric field strength?

(A) $(6.7 \times 10^{-7})\mathbf{a}_r$ N/C

(B) $(2.0 \times 10^{-4})\mathbf{a}_r$ N/C

(C) $(7.5 \times 10^{4})\mathbf{a}_r$ N/C

(D) $(2.3 \times 10^{5})\mathbf{a}_r$ N/C

3. The following values for relative permittivity are taken from a standard reference: paper, 2.0; quartz, 5.0; and marble, 8.3.

The electric field strength is 2.0×10^{5} V/m. $\mathbf{D}$ must be greater than or equal to 7×10^{-6} C/m^2. Determine which of the given dielectrics is suitable.

4. Moist soil has a conductivity of approximately 10^{-3} S/m and an ϵ_r of 3.0. A certain electric field strength is given by

$$E = 3.5 \times 10^{-6} \sin(2.5 \times 10^{3})t \ \frac{\text{V}}{\text{m}}$$

Determine the displacement current density.

5. In which circuit element is displacement current largest: conductor, inductor, or capacitor? Explain.

SOLUTIONS

1. The polarization for a material with susceptibility χ_e is

$$P = \chi_e \epsilon_0 E$$

This assumes that the electric field $\mathbf{E}$ and the polarization $\mathbf{P}$ are in the same direction; that is, the material is isotropic and linear. Given that, the electic flux density is

$$D = \epsilon_0 E + P$$

Substituting P/χ_e for $\epsilon_0 E$,

$$D = \frac{P}{\chi_e} + P$$

$$= \frac{2.0 \times 10^{-6} \ \dfrac{\text{C}}{\text{m}^2}}{2.2} + 2.0 \times 10^{-6} \ \frac{\text{C}}{\text{m}^2}$$

$$= \boxed{2.9 \times 10^{-6} \ \text{C/m}^2}$$

The answer is (C).

2. The flux density $\mathbf{D}$ in a medium of permittivity ϵ_r is

$$\mathbf{D} = \epsilon_0 \epsilon_r \mathbf{E}$$

Rearranging gives

$$\mathbf{E} = \frac{\mathbf{D}}{\epsilon_0 \epsilon_r} = \frac{(2 \times 10^{-6})\mathbf{a}_r \ \dfrac{\text{C}}{\text{m}^2}}{\left(8.854 \times 10^{-12} \ \dfrac{\text{C}^2}{\text{N} \cdot \text{m}^2}\right)(3.0)}$$

$$= \boxed{(7.5 \times 10^{4})\mathbf{a}_r \ \text{N/C}}$$

The answer is (C).

3. Determine the minimum value of the permittivity, ϵ_r, required.

$$\mathbf{D} = \epsilon_0 \epsilon_r \mathbf{E}$$

Assuming the dielectrics to be isotropic and linear,

$$D = \epsilon_0 \epsilon_r E$$

$$\epsilon_r = \frac{D}{E \epsilon_0} = \frac{7 \times 10^{-6} \ \dfrac{\text{C}}{\text{m}^2}}{\left(2.0 \times 10^{5} \ \dfrac{\text{V}}{\text{m}}\right)\left(8.854 \times 10^{-12} \ \dfrac{\text{C}^2}{\text{N} \cdot \text{m}^2}\right)}$$

$$= 3.95$$

Therefore, the quartz or marble is an acceptable dielectric.

4. Determine the value of the magnitude of the electric flux density, D.

$$D = \epsilon_0 \epsilon_r E$$

$$= \left(8.854 \times 10^{-12} \; \frac{C^2}{N \cdot m^2}\right)(3.0)$$

$$\times \left(3.5 \times 10^{-6} \sin(2.5 \times 10^3)t \; \frac{V}{m}\right)$$

$$= 9.30 \times 10^{-17} \sin(2.5 \times 10^3)t \; C/m^2$$

The displacement current is

$$J_d = \frac{\partial D}{\partial t}$$

$$= \frac{\partial}{\partial t}\left((9.30 \times 10^{-17})\sin(2.5 \times 10^3)t \; \frac{C}{m^2}\right)$$

$$= (9.3 \times 10^{-17})\left((2.5 \times 10^3)\cos(2.5 \times 10^3)t \; \frac{A}{m^2}\right)$$

$$= \boxed{(2.32 \times 10^{-13})\cos(2.5 \times 10^3)t \; A/m^2}$$

5. The displacement current is largest in a capacitor. The charge stored as a capacitor, of capacitance C, is

$$Q = CV$$

The capacitance is proportional to the area.

$$C \propto A$$

Capacitors, in general, have large areas and store high concentrations of charges. Therefore, the electric field strength is large and, consequently, so is $\mathbf{D}$. At high frequencies this results in a significant displacement current.

22 Magnetostatics

PRACTICE PROBLEMS

1. The flux density a distance r from a current-carrying conductor of length dl varies as

(A) $\ln r$

(B) $1/r$

(C) $1/r^2$

(D) constant

2. Determine the magnitude of the core flux density for one turn of wire wrapped around a cast-iron core with a relative permeability, μ_r, of 400. The wire is rated at 15 A. The core measures $0.10 \text{ m} \times 0.15 \text{ m}$.

3. In a straight, infinitely long conductor, how much current does it take to produce a magnetic field 1 m away that is equivalent in strength to the Earth's, that is, $5 \times 10^{-5} \text{ T}$?

4. Consider the following illustration.

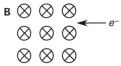

The magnetic field, **B**, is uniform and directed into the paper. An electron is injected into the field from the right with some initial velocity. In what direction does the electron initially deflect?

(A) up

(B) down

(C) left

(D) right

5. A galvanometer moving-coil, used in instrumentation, is placed between two permanent magnets with a **B**-field strength of $4.0 \times 10^{-3} \text{ T}$ in the directions shown. The radius of the coil is 0.005 m. The coil contains 10 turns, or loops, and has a current rating of 0.001 A. Determine the maximum torque that the meter can generate against the restoring spring.

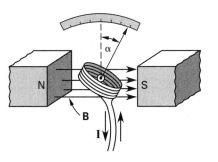

SOLUTIONS

1. The Biot-Savart law gives the magnitude of flux density in differential form as

$$dB = \left(\frac{\mu_0 I \, dl}{4\pi r^2}\right) \sin\theta$$

The answer is (C).

2. The flux produced by a current I that travels N turns over a distance l is

$$\Phi = \mu N I l = BA$$

Rearranging to solve for the flux density B over area A gives

$$B = \frac{\mu N I l}{A} = \frac{\mu_0 \mu_r N I l}{A}$$

$$= \frac{\left(1.2566 \times 10^{-6} \ \frac{\text{N}}{\text{A}^2}\right)(400)(1)(15 \ \text{A})}{\times (0.10 \ \text{m} + 0.15 \ \text{m} + 0.10 \ \text{m} + 0.15 \ \text{m})}$$

$$= \boxed{0.25 \ \text{T}}$$

3. The magnetic field, B, of a straight, infinitely long conductor is

$$B = \frac{\mu I}{2\pi r}$$

Because no specific condition was given, free space permeability, μ_0, is assumed. Rearranging and substituting gives

$$I = \frac{2\pi r B}{\mu_0}$$

$$= \frac{2\pi \, (1 \ \text{m})(5 \times 10^{-5} \ \text{T})}{1.2566 \times 10^{-6} \ \frac{\text{H}}{\text{m}}}$$

$$= \boxed{250 \ \text{A}}$$

4. The magnetic force on a moving charged particle q, moving with velocity $\mathbf{v}$ in magnetic field $\mathbf{B}$ is

$$\mathbf{F} = q\mathbf{v} \times \mathbf{B}$$

Using the right-hand rule to determine the direction of the cross-product, the initial deflection appears to be down toward the bottom of the paper. But the electron's charge is negative. Therefore, the deflection is upward toward the top of the paper.

The answer is (A).

5. The torque on magnetic moment $\mathbf{m}$ in field $\mathbf{B}$ is

$$\mathbf{T} = \mathbf{m} \times \mathbf{B}$$

The strength of the $\mathbf{B}$-field is given. The magnetic moment, m, can be determined from

$$m = IA$$

Because there are N turns of wire, each carrying a current I, the equation is

$$m = NIA = NI\pi r^2$$

$$= (10)(0.001 \ \text{A})(\pi)(0.005 \ \text{m})^2$$

$$= 7.85 \times 10^{-7} \ \text{A·m}^2$$

The maximum magnetic torque can now be determined by substitution.

$$\mathbf{T} = \mathbf{m} \times \mathbf{B}$$

$$T_{\text{max}} = mB \sin\frac{\pi}{2} = mB$$

$$= (7.85 \times 10^{-7} \ \text{A·m}^2)(4.0 \times 10^{-3} \ \text{T})$$

$$= \boxed{3.1 \times 10^{-9} \ \text{N·m}}$$

23 Magnetostatic Fields

PRACTICE PROBLEMS

1. Define magnetization, and state its mathematical formula with appropriate units.

2. Write the formula for the magnetic flux density that is always valid regardless of the linearity of the material. Use the SI form of the equation.

3. By what factor does the flux density increase if high-purity iron is substituted for commercial iron as the core of a transformer?

   (A) 4

   (B) 30

   (C) 6000

   (D) 2×10^5

4. The magnetic field strength of Earth is approximately 40 A/m. Determine the current required in a single turn of wire located at the equator to generate the same magnetic field at distances far from Earth's surface.

5. What type of magnetic material generates an internal magnetic field that opposes the applied field?

   (A) diamagnetic

   (B) paramagnetic

   (C) ferromagnetic

   (D) antiferromagnetic

SOLUTIONS

1. Magnetization, $\mathbf{M}$, is the magnetic moment per unit volume, and is given by

$$\mathbf{M} = \frac{\mathbf{m}}{v}$$

The units are A/m.

2. The formula gives the magnetic flux density, $\mathbf{B}$, in terms of the magnetic field, $\mathbf{H}$, and the magnetization, $\mathbf{M}$, in the permittivity of free space, μ_0.

$$\boxed{\mathbf{B} = \mu_0\mathbf{H} + \mu_0\mathbf{M}}$$

3. The relative permeability, μ_r, of commercial-grade iron is approximately 6000, while that of high-purity iron is 2×10^5. The flux density magnitude is

$$B = \mu_0\mu_r H$$

The only change from one material to the next is μ_r. Thus, the factor f is

$$f = \frac{\mu_r \text{ (high purity)}}{\mu_r \text{ (commercial)}}$$

$$= \frac{2 \times 10^5}{6000}$$

$$= \boxed{33 \quad (30)}$$

The answer is (B).

4. At distances far from Earth's surface, the applicable equation is

$$H = \frac{NI}{l}$$

Rearranging gives

$$I = \frac{Hl}{N}$$

Earth's radius is approximately 6.4×10^6 m.

The path length is

$$l = 2\pi r = 2\pi(6.4 \times 10^6 \text{ m})$$

$$= 4.0 \times 10^7 \text{ m}$$

Substituting the known and calculated information gives

$$I = \frac{Hl}{N} = \frac{\left(40 \ \frac{A}{m}\right)(4.0 \times 10^7 \ m)}{1}$$
$$= \boxed{1.6 \times 10^9 \ A}$$

5. Diamagnetic materials tend to reduce the flux by generating an opposing internal magnetic field.

The answer is (A).

24 Electrodynamics

PRACTICE PROBLEMS

1. The following information is taken from a certain portable generator: v is 1450 m/s, B is 1.2×10^{-2} T, loop length is 0.4 m, and output voltage is 120 V. Estimate the number of turns required to induce the indicated voltage.

(A) 1.0

(B) 17

(C) 18

(D) 120

2. Which law describes induced electromotance?

(A) Faraday's law

(B) Ampere's law

(C) Coulomb's law

(D) Gauss' law

3. Which of the following terms represents the magnetomotive force, or mmf?

(A) $\oint \mathbf{E} \cdot d\mathbf{l}$

(B) $\oint \epsilon \mathbf{D} \cdot d\mathbf{l}$

(C) $\mu_0 \mathbf{H}$

(D) $\oint \mathbf{H} \cdot d\mathbf{l}$

4. In the following illustration, determine the direction of energy flow.

5. What fundamental action must occur to a charge in order to produce an electromagnetic wave capable of transporting energy in free space?

SOLUTIONS

1. The induced voltage resulting from a conductor moving in a uniform magnetic field is

$$v = -NBl\mathrm{v}$$

Rearranging and solving gives

$$N = \left| \frac{v}{Bl\mathrm{v}} \right|$$

$$= \frac{120 \text{ V}}{(1.2 \times 10^{-2} \text{ T})(0.4 \text{ m})\left(1450 \, \frac{\text{m}}{\text{s}}\right)}$$

$$= 17.2$$

Therefore, 18 turns are required as a minimum.

The answer is (C).

2. Faraday's law of magnetic induction is used to determine the amount of induced voltage, also called induced electromotance.

The answer is (A).

3. The mmf is the integral of the magnetic field strength, $\mathbf{H}$, along the path, $d\mathbf{l}$, where the magnetic flux is present. It is often given as NI, the number of turns times the current. The force produced per unit flux set up in the conductor is $\mathbf{H}$, hence the name.

The answer is (D).

4. The direction of energy flow, or power density, is given by Poynting's vector, $\mathbf{S}$.

$$\mathbf{S} = \mathbf{E} \times \mathbf{H}$$

In terms of the $\mathbf{B}$-field, assuming a linear, isotropic medium,

$$\mathbf{S} = \frac{1}{\mu} \mathbf{E} \times \mathbf{B}$$

If the material were not isotropic, the permeability would remain with the $\mathbf{B}$-field as part of the cross product. The direction of $\mathbf{S}$ would then depend on the product of the permeability and magnetic flux density fields, $\mu\mathbf{B}$, as well.

Using the right-hand rule, the direction of the cross product **S** is into the plane of the paper, $\otimes$.

5. The charge must be accelerated or decelerated by some means.

A stationary charge is related to an electric field, a charge in motion with uniform velocity is related to a magnetic field, and an accelerating charge is related to a radiated field.

Field Theory

25 Maxwell's Equations

PRACTICE PROBLEMS

1. Which of the following point forms of Maxwell's equations implies the nonexistence of magnetic monopoles?

(A) $\nabla \cdot \mathbf{D} = \rho$

(B) $\nabla \cdot \mathbf{B} = 0$

(C) $\nabla \times \mathbf{E} = -\dfrac{\partial \mathbf{B}}{\partial t}$

(D) $\nabla \cdot \mathbf{D} = 0$

2. What is missing from the following equation?

$$\nabla \times \mathbf{H} = \mathbf{J}_c + \dfrac{\partial}{\partial t}$$

(A) **B**

(B) **D**

(C) **E**

(D) **H**

3. Which of the following equations for magnetic circuits is analogous to the equation shown for an electric circuit?

$$\mathrm{emf} = V = IR$$

(A) $G = \dfrac{1}{R}$

(B) $P_m = \dfrac{\mu A}{l}$

(C) $\mathcal{R} = \dfrac{l}{\mu A}$

(D) $V_m = \phi \mathcal{R}$

4. Which of the following is the continuity equation?

(A) $\nabla \cdot \mathbf{D} = \rho$

(B) $\mathbf{J} = \sigma \mathbf{E}$

(C) $\nabla \cdot \mathbf{J} = -\dfrac{\partial \rho}{\partial t}$

(D) $\mathbf{F} = Q(\mathbf{E} + \mathbf{v} + \mathbf{B})$

5. What is missing in the following equation?

$$\nabla \times \mathbf{E} = -\dfrac{\partial}{\partial t}$$

(A) **B**

(B) **D**

(C) **H**

(D) **J**

SOLUTIONS

1. The nonexistence of magnetic monopoles is implied by the divergence of the magnetic flux density, **B**, being equal to zero.

$$\boxed{\nabla \cdot \mathbf{B} = 0}$$

The answer is (B).

2.
$$\nabla \times \mathbf{H} = \mathbf{J}_c + \mathbf{J}_d$$
$$\mathbf{J}_d = \frac{\partial \mathbf{D}}{\partial t}$$

$$\boxed{\text{The electric flux density, } \mathbf{D}, \text{ is missing from the displacement current term.}}$$

The answer is (B).

3. The magnetomotive force (mmf) is analogous to the electromotive force (emf). The analogous magnetic equation is

$$\boxed{\text{mmf} = V_m = \phi \mathcal{R}}$$

The answer is (D).

4. The continuity equation states that the divergence of the current density equals the negative rate of change of the charge density.

$$\boxed{\nabla \cdot \mathbf{J} = -\frac{\partial \rho}{\partial t}}$$

The answer is (C).

5. The curl of the electric field is equal to the negative rate of change of the magnetic flux density.

$$\nabla \times \mathbf{E} = -\frac{\partial \mathbf{B}}{\partial t}$$

$$\boxed{\text{The magnetic flux density term, } \mathbf{B}, \text{ is missing.}}$$

The answer is (A).

Field Theory

26 DC Circuit Fundamentals

PRACTICE PROBLEMS

1. What is the power dissipated in R_3 of the circuit shown?

2. Calculate the resistance of the load necessary for maximum power transfer to occur in the circuit shown.

3. For the circuit shown, which method will result in fewer equations: the loop-current method or the node-voltage method?

4. Using the loop-current method, solve for the current in R_3 in Prob. 3.

5. Using the node-voltage method, solve for the current in R_3 in Prob. 3.

SOLUTIONS

1. Power is given by

$$P = VI = I^2 R = \frac{V^2}{R}$$

Using the last form of the equation, determine the voltage across R_3. Combine the parallel resistors R_2 and R_3.

$$R_{2/3} = \frac{R_2 R_3}{R_2 + R_3} = \frac{(10\ \Omega)(10\ \Omega)}{10\ \Omega + 10\ \Omega} = 5\ \Omega$$

The circuit is transformed as follows.

Using the voltage-divider concept, the voltage across $R_{2/3}$ (and R_3) is

$$V_{2/3} = V_3 = (50\ \text{V})\left(\frac{R_{2/3}}{R_1 + R_{2/3}}\right)$$

$$= (50\ \text{V})\left(\frac{5\ \Omega}{5\ \Omega + 5\ \Omega}\right)$$

$$= 25\ \text{V}$$

The power dissipated in R_3 is

$$P_3 = \frac{V_3^2}{R_3} = \frac{(25\ \text{V})^2}{10\ \Omega}$$

$$= \boxed{62\ \text{W}}$$

2. Maximum power transfer occurs when the load resistance equals the Thevenin or Norton equivalent resistance.

Open circuit the current source and determine R_{Th}.

The 15 Ω and 20 Ω resistors are in an open-circuit path.

$$R_{\text{Th}} = R_3 = \boxed{5 \ \Omega}$$

3. The loop-current method using Kirchoff's voltage law (KVL) uses $n-1$ equations, where n is the number of loops. There are three loops present. The number of equations required is two.

The node-voltage method using Kirchoff's current law (KCL) requires $n-1$ equations, where n is the number of principal nodes. There are two principal nodes. Only one equation is required.

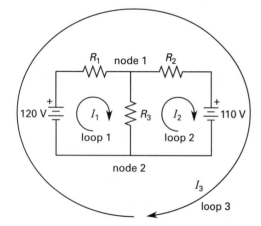

The node-voltage method requires fewer equations.

4. Consider the following circuit.

Using the selected loops with polarities for the resistive elements based on the loop-current direction, KVL for loop 1 is

$$120 \text{ V} - I_1 R_1 - (I_1 - I_2)R_3 = 0$$

Rearranging gives

$$I_1(R_1 + R_3) - I_2 R_3 = 120 \text{ V} \qquad \text{[I]}$$

KVL for loop 2 is

$$-(I_2 - I_1)R_3 - I_2 R_2 - 110 \text{ V} = 0$$

Rearranging gives

$$I_1 R_3 - I_2(R_3 + R_2) = 110 \text{ V} \qquad \text{[II]}$$

There are now two equations, [I] and [II], with two unknowns, I_1 and I_2. Use Cramer's rule and matrix algebra to solve for I_1 and I_2.

$$\begin{vmatrix} R_1 + R_3 & -R_3 \\ R_3 & -(R_3 + R_2) \end{vmatrix} \begin{vmatrix} I_1 \\ I_2 \end{vmatrix} = \begin{vmatrix} 120 \text{ A} \\ 110 \text{ A} \end{vmatrix}$$

$$\begin{vmatrix} 35 \text{ A} & -30 \text{ A} \\ 30 \text{ A} & -40 \text{ A} \end{vmatrix} \begin{vmatrix} I_1 \\ I_2 \end{vmatrix} = \begin{vmatrix} 120 \text{ A} \\ 110 \text{ A} \end{vmatrix}$$

$$I_1 = \frac{\begin{vmatrix} 120 \text{ A} & -30 \text{ A} \\ 110 \text{ A} & -40 \text{ A} \end{vmatrix}}{\begin{vmatrix} 35 \text{ A} & -30 \text{ A} \\ 30 \text{ A} & -40 \text{ A} \end{vmatrix}} = \frac{-1500 \text{ A}}{-500 \text{ A}} = 3 \text{ A}$$

$$I_2 = \frac{\begin{vmatrix} 35 \text{ A} & 120 \text{ A} \\ 30 \text{ A} & 110 \text{ A} \end{vmatrix}}{\begin{vmatrix} 35 \text{ A} & -30 \text{ A} \\ 30 \text{ A} & -40 \text{ A} \end{vmatrix}} = \frac{250 \text{ A}}{-500 \text{ A}} = -0.5 \text{ A}$$

Summing to obtain the current through R_3 and noting that the direction of I_2 was incorrect gives

$$I_3 = |I_1| + |I_2| = 3 \text{ A} + 0.5 \text{ A}$$
$$= \boxed{3.5 \text{ A}}$$

The direction of the current I_3 is positive and identical to the direction of I_1.

5. Consider the following circuit.

The principal node A is labeled. The node at the lower portion of the circuit is chosen as the reference or ground node. All currents are assumed to be positive.

Writing KCL for node A gives

$$\frac{V_A - 120 \text{ V}}{R_1} + \frac{V_A - 0 \text{ V}}{R_3} + \frac{V_A - 110 \text{ V}}{R_2} = 0$$

Solve for V_A.

$$\frac{V_A}{R_1} - \frac{120 \text{ V}}{R_1} + \frac{V_A}{R_3} + \frac{V_A}{R_2} - \frac{110 \text{ V}}{R_2} = 0$$

$$\frac{V_A}{R_1} + \frac{V_A}{R_3} + \frac{V_A}{R_2} = \frac{120 \text{ V}}{R_1} + \frac{110 \text{ V}}{R_2}$$

$$V_A = \frac{\dfrac{120 \text{ V}}{R_1} + \dfrac{110 \text{ V}}{R_2}}{\dfrac{1}{R_1} + \dfrac{1}{R_3} + \dfrac{1}{R_2}}$$

$$= \frac{\dfrac{120 \text{ V}}{5 \text{ }\Omega} + \dfrac{110 \text{ V}}{10 \text{ }\Omega}}{\dfrac{1}{5 \text{ }\Omega} + \dfrac{1}{30 \text{ }\Omega} + \dfrac{1}{10 \text{ }\Omega}}$$

$$= 105 \text{ V}$$

Use V_A to determine the desired current.

$$I_3 = \frac{V_A}{R_3} = \frac{105 \text{ V}}{30 \text{ }\Omega} = \boxed{3.5 \text{ A}}$$

Circuit Theory

27 AC Circuit Fundamentals

PRACTICE PROBLEMS

1. Express the phasor voltage 120 V∠30° as a function of time if the frequency is 60 Hz. The voltage is given in "effective value phasor" notation.

2. Express each of the following in rectangular form.

(a) $15\angle -180°$

(b) $10\angle 37°$

(c) $50\angle 120°$

(d) $21\angle -90°$

3. Express each of the following in phasor form.

(a) $6 + j7$

(b) $50 - j60$

(c) $-75 + j45$

(d) $90 - j180$

4. Calculate the following, and express each in phasor form.

(a) $(7 + j5)/6$

(b) $(13 + j17)/(15 - j10)$

(c) $(0.020\angle 90°)/0.034\angle 56°$

5. A sinusoidal waveform has a peak value at $t = 1$ ms and a period of 10 ms. Express this signal as a cosine and sine function.

6. A voltage of $(10\text{ V})\cos(100t + 25°)$ is applied to a resistance of 25 Ω and an inductance of 0.50 H in parallel. What are the (a) phase angle, and (b) rms value of the sum of the currents?

7. Draw the power triangle for the circuit in Prob. 6.

8. What is the capacitance, in units of farads, to completely correct the power factor of the circuit in Prob. 6, that is, to make $Q = 0$?

9. An 80 μF capacitor is in series with a 9 mH inductor. (a) What is the total impedance (at 60 Hz)? (b) Is the circuit considered a leading or lagging circuit?

10. If the capacitor and inductor in Prob. 9 are connected in parallel, what is the total impedance, and is the circuit considered leading or lagging?

SOLUTIONS

1. As a function of time, t, with peak voltage V_p and angular frequency ω,

$$v(t) = V_p \sin(\omega t + \theta)$$

The peak voltage is proportional to the effective voltage, V_{eff},

$$\begin{aligned} V_p &= V_{\text{eff}}\sqrt{2} \\ &= 120 \text{ V}\sqrt{2} \\ &= 169.71 \text{ V} \end{aligned}$$

The angular frequency is

$$\begin{aligned} \omega &= 2\pi f = 2\pi(60 \text{ Hz}) \\ &= 376.99 \text{ rad/s} \end{aligned}$$

The given angle of $30°$ equals $\pi/6$ rad, so

$$v(t) = \boxed{170 \sin\left(377t + \frac{\pi}{6}\right)}$$

2. (a) $\quad 15\cos(-180°) + j15\sin(-180°)$

$$= \boxed{-15 + j0}$$

(b) $\quad 10\cos 37° + j10\sin 37°$

$$= \boxed{7.99 + j6.02}$$

(c) $\quad 50\cos 120° + j50\sin 120°$

$$= \boxed{-25 + j43.3}$$

(d) $\quad 21\cos(-90°) + j21\sin(-90°)$

$$= \boxed{0 - j21}$$

3. (a) $\quad \sqrt{(6)^2 + (7)^2}\angle\tan^{-1}\dfrac{7}{6}$

$$= \boxed{9.22\angle 49.4°}$$

(b) $\quad \sqrt{(50)^2 + (-60)^2}\angle\tan^{-1}\dfrac{-60}{50}$

$$= \boxed{78.10\angle -50.19°}$$

(c) $\quad \sqrt{(-75)^2 + (45)^2}\angle\tan^{-1}\dfrac{45}{-75}$

$$= \boxed{87.46\angle 149.04°}$$

(d) $\quad \sqrt{(90)^2 + (-180)^2}\angle\tan^{-1}\dfrac{-180}{90}$

$$= \boxed{201.25\angle -63.43°}$$

4. (a) $\quad \dfrac{1}{6}\sqrt{(7)^2 + (5)^2}\angle\tan^{-1}\dfrac{5}{7}$

$$= \boxed{1.4\angle 35.54°}$$

(b) $\quad \dfrac{\sqrt{(13)^2 + (17)^2}\,\tan^{-1}\dfrac{17}{13}}{\sqrt{(15)^2 + (-10)^2}\,\tan^{-1}\dfrac{-10}{15}}$

$$= \frac{21.4\angle 52.59°}{18.0\angle -33.69°}$$

$$= \boxed{1.19\angle 86.28°}$$

(c) $\qquad\qquad \boxed{0.588\angle 34°}$

5.

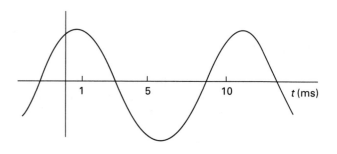

$$\omega = \frac{2\pi}{T} = \frac{2\pi}{0.01} = 200\pi \text{ rad/s}$$

$$10 \text{ ms} = 360°$$

$$1 \text{ ms} = 36°$$

For the sine, the positive zero crossing is at -1.5 ms, or $-54°$.

$$f = \boxed{F_p \sin(200\pi t + 54°) \text{ or } F_p \sin(200\pi t + 0.9)}$$
$$= \boxed{F_p \cos(200\pi t - 36°) \text{ or } F_p \cos(200\pi t - 0.6)}$$

6. Using a cosine reference,

(a)
$$I_R = \frac{10 \text{ V} \angle 25°}{25 \ \Omega} = 0.4 \text{ A} \angle 25°$$
$$= 0.3625 + j0.1690 \text{ A}$$
$$I_L = \frac{10 \text{ V} \angle 25°}{(0.5)(100) \angle 90° \ \Omega}$$
$$= 0.2 \text{ A} \angle -65°$$
$$= 0.0845 - j0.1813 \text{ A}$$
$$I_t = 0.4470 - j0.0123 \text{ A}$$
$$= 0.4472 \text{ A} \angle -1.6°$$
$$\text{phase angle} = \boxed{-1.6°}$$

(b) The calculations were made using peak values. Divide the total peak current by $\sqrt{2}$ to obtain the rms values.

$$I_{\text{rms}} = \frac{0.4472 \text{ A}}{\sqrt{2}} = \boxed{0.316 \text{ A}}$$

7. Use rms values. The magnitude, S, of the complex power vector is the product of the voltage, V, and the magnitude of the complex conjugate of the current, I^*.

$$S = VI^* = \frac{(10 \text{ V} \angle 25°)(0.4472 \text{ A} \angle 1.6°)}{\sqrt{2}\sqrt{2}}$$
$$= 2.236 \angle 26.6° \text{ VA}$$
$$= 2 + j1 \text{ VA}$$

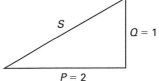

8. The capacitance current must cancel the inductance current, so its reactance must be $-j\omega L = -j50 \ \Omega$.

$$X_C = \frac{1}{j\omega C} = \frac{-j}{100 C} \rightarrow -j50 \ \Omega$$
$$C = \frac{1}{(50 \ \Omega)\left(100 \ \dfrac{\text{rad}}{\text{s}}\right)}$$
$$= \boxed{200 \times 10^{-6} \text{ F} \quad (200 \ \mu\text{F})}$$

9. The impedance, Z_C, for the capacitor is

$$Z_C = \frac{1}{j\omega C} = \frac{1}{j\left(377 \ \dfrac{\text{rad}}{\text{s}}\right)(80 \times 10^{-6} \text{ F})}$$
$$= -j(33.16 \ \Omega)$$
$$= 33.16 \ \Omega \angle -90°$$

The impedance, Z_L, for the inductor is

$$Z_L = j\omega L = j\left(377 \ \dfrac{\text{rad}}{\text{s}}\right)(9 \times 10^{-3} \text{ H})$$
$$= j3.39 \ \Omega$$
$$= 3.39 \ \Omega \angle 90°$$

For a series circuit, the impedance is

$$Z_{\text{total}} = Z_C + Z_L$$
$$= -j33.16 \ \Omega + j3.39 \ \Omega$$
$$= -j29.77 \ \Omega$$
$$= \boxed{29.77 \ \Omega \angle -90°}$$

The impedance angle is negative. The circuit is capacitative and therefore leading.

10. For a parallel circuit, the impedance is

$$\frac{1}{Z_{\text{total}}} = \frac{1}{Z_C} + \frac{1}{Z_L}$$
$$= \frac{1}{33.16 \ \Omega \angle -90°} + \frac{1}{3.39 \ \Omega \angle 90°}$$
$$= 0.0302 \ \Omega \angle 90° + 0.2950 \ \Omega \angle -90°$$
$$= j0.0302 \ \Omega - j0.2950 \ \Omega$$
$$= -j0.2648 \ \Omega$$

Circuit Theory

So,

$$Z_{\text{total}} = \frac{1}{-j0.2648 \ \Omega} = \frac{j}{0.2648 \ \Omega}$$
$$= \boxed{3.78 \ \Omega \angle 90°}$$

The impedance angle is positive. The circuit is inductive and, therefore, lagging.

28 Transformers

PRACTICE PROBLEMS

1. Which of the following describes the leakage flux?

(A) flux outside the core

(B) flux linking the primary

(C) flux linking the secondary

(D) flux other than the mutual flux

2. For the transformer shown, which of the following statements is true?

(A) $M > 0$

(B) $M < 0$

(C) $X_m > 0$ and $X_p > 0$

(D) Both (A) and (C) are true.

3. An ideal transformer has 90 turns on the primary and 2200 turns on the secondary. The load draws 8 A at 0.8 pf. What is the current in the primary?

(A) 90 A∠37°

(B) 196 A∠37°

(C) 320 A∠1°

(D) 333 A∠0.2π

4. An ideal transformer with a 120 V, 60 Hz source has a turns ratio of 10. What is the peak flux in the secondary?

5. A step-down transformer consists of 200 primary turns and 40 secondary turns. The primary voltage is 550 V. If an impedance of 4.2 Ω is the secondary load, what are the (a) secondary voltage, (b) primary current, and (c) secondary current?

6. For the circuit shown, if the amplifier and load impedances are matched, what is the turns ratio?

7. A 100 V (rms) source with an internal resistance of 8 Ω is connected to a 200/100-turn step-down transformer. An impedance of 6 Ω + $j8$ is connected to the secondary. Find the (a) primary current and (b) secondary power.

SOLUTIONS

1. Leakage flux is defined as flux that goes beyond its intended path and does not serve its intended purpose. That is, it is the flux that links only one coil or the other, but not both, in a two-coil transformer. The flux linking both coils is the mutual flux.

The answer is (D).

2. Both assumed currents enter at the dotted terminals. Therefore, the mutual inductor is positive. Further, the mutual reactance, X_m, and the self-inductance of the coils, X_p (or X_s), are the same sign.

The answer is (D).

3. The turns ratio a for a transformer with N windings and currents I (p, primary; s, secondary) is

$$a = \frac{N_p}{N_s} = \frac{I_s}{I_p}$$

Therefore,

$$a = \frac{N_p}{N_s} = \frac{90}{2200} = 0.04091$$

$$= \frac{I_s}{I_p}$$

The magnitude of the current is

$$I_p = \frac{I_s}{a} = \frac{8 \text{ A}}{0.04091} = 195.6 \text{ A}$$

In an ideal transformer, no phase shift occurs, so the power factor is

$$\text{pf} = \cos\phi$$
$$0.8 = \cos\phi$$
$$\phi = \cos^{-1} 0.8 = 36.87°$$

Therefore,

$$\boxed{\mathbf{I}_p = 196 \text{ A}\angle 37°}$$

The answer is (B).

4. The effective voltage, $V_s(t)$, that is induced in the secondary within a transformer with mutual flux Φ_m, is

$$v_s(t) = -4.44 N_s f \Phi_m \cos\omega t$$

Because only the peak flux was requested,

$$V_s = 4.44 N_s f \Phi_m$$

$$\Phi_m = \frac{V_s}{4.44 N_s f}$$

Now, from the turns ratio,

$$a = \frac{N_p}{N_s} = \frac{V_p}{V_s}$$

$$V_s = \left(\frac{N_s}{N_p}\right) V_p = \left(\tfrac{1}{10}\right)(120 \text{ V}) = 12 \text{ V}$$

Substituting gives

$$\Phi_m = \frac{V_s}{4.44 N_s f} = \frac{12 \text{ V}}{(4.44)(1)(60 \text{ Hz})}$$

$$= 4.54 \times 10^{-2} \text{ Wb} \quad (45 \text{ mWb})$$

(This is the same flux present in the primary.)

5. (a) $\quad V_s = \left(\dfrac{N_s}{N_p}\right) V_p = \left(\dfrac{40}{200}\right)(550 \text{ V})$

$$= \boxed{110 \text{ V}}$$

(b) $\quad I_p = \dfrac{V_s^2}{V_p Z_s} = \dfrac{(110 \text{ V})^2}{(550 \text{ V})(4.2 \text{ }\Omega)}$

$$= \boxed{5.24 \text{ A}}$$

(c) $\quad I_s = \dfrac{V_s}{Z_s} = \dfrac{110 \text{ V}}{4.2 \text{ }\Omega}$

$$= \boxed{26.2 \text{ A}}$$

6. $\quad \dfrac{N_p}{N_s} = \sqrt{\dfrac{Z_p}{Z_s}} = \sqrt{\dfrac{200 \text{ }\Omega}{\dfrac{8 \text{ }\Omega}{3}}}$

$$= \boxed{8.66}$$

7. Consider the following circuit.

The circuit can be simplified to

$$\mathbf{Z}_p = \left(\frac{N_p}{N_s}\right)^2 \mathbf{Z}_s = \left(\frac{200}{100}\right)^2 (10 \ \Omega \angle 53.13°)$$

$$= 40 \ \Omega \angle 53.13°$$

$$= (24 + j32) \ \Omega$$

The circuit can be further simplified to

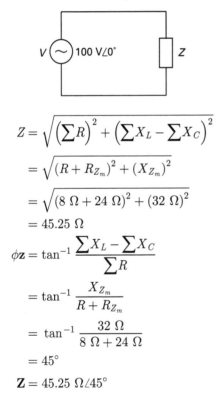

$$Z = \sqrt{\left(\sum R\right)^2 + \left(\sum X_L - \sum X_C\right)^2}$$

$$= \sqrt{(R + R_{Z_m})^2 + (X_{Z_m})^2}$$

$$= \sqrt{(8 \ \Omega + 24 \ \Omega)^2 + (32 \ \Omega)^2}$$

$$= 45.25 \ \Omega$$

$$\phi\mathbf{z} = \tan^{-1} \frac{\sum X_L - \sum X_C}{\sum R}$$

$$= \tan^{-1} \frac{X_{Z_m}}{R + R_{Z_m}}$$

$$= \tan^{-1} \frac{32 \ \Omega}{8 \ \Omega + 24 \ \Omega}$$

$$= 45°$$

$$\mathbf{Z} = 45.25 \ \Omega \angle 45°$$

(a)
$$\mathbf{I}_p = \frac{\mathbf{V}}{\mathbf{Z}} = \frac{100 \ \text{V} \angle 0°}{45.25 \ \Omega \angle 45°}$$

$$= \boxed{2.21 \ \text{A}_{\text{rms}} \angle -45°}$$

(b)
$$P_s = I_s^2 R_s = I_s^2 Z_s \cos\phi$$

$$= \left(\frac{N_p I_p}{N_s}\right)^2 Z_s \cos\phi$$

$$= \left(\frac{(200)(2.21 \ \text{A})}{100}\right)^2 (10 \ \Omega) \cos 53.13°$$

$$= \boxed{117.2 \ \text{W}}$$

Circuit Theory

29 Linear Circuit Analysis

PRACTICE PROBLEMS

1. For which of the following circuit elements is super-position valid?

    (A) independent source

    (B) dependent source

    (C) charged capacitor

    (D) inductor with $\mathbf{I}_{\text{initial}} \neq 0$

2. A 5 Ω length of copper wire is used in a circuit designed for a maximum temperature of 50°C. What is the resistance of the wire at maximum design temperature specification?

3. Two square parallel plates (0.04 m × 0.04 m) are separated by a 0.1 cm thick insulator with a dielectric constant of 3.4. The plates are connected across 200 V. What is the capacitance?

4. For the parallel plates described in Prob. 3, what charge exists on the plates?

5. An 8 mH inductor carries 15 A of current. What is the average magnetic flux present?

6. For the circuit shown, which resistor(s) is (are) redundant?

    (A) R_1 only

    (B) R_1 and R_3

    (C) R_2 and R_4

    (D) R_1 and R_4

7. For the circuit shown, what current is flowing in the 8 Ω resistor?

8. For the circuit shown, what current is flowing in the 6 Ω resistor?

9. At $t=0$, the switch connects to position A. At $t=5 \times 10^{-4}$ s, the switch is moved to position B. Assume there is no current anywhere in the circuit prior to $t=0$. (a) What is the current flowing in the circuit just prior to the switch moving to position B? (b) What is the equation of the current after the switch is moved to position B?

10. For the circuit shown, find the impedance, z, parameters.

SOLUTIONS

1. Capacitors and inductors are linear elements for which superposition would be valid only if they were initially uncharged or without an initial magnetic field, respectively. Therefore, neither (C) nor (D) is the answer. The output of dependent sources depends on electrical parameters elsewhere in the circuit and is therefore not linear. Therefore, (B) is not the answer.

Independent sources are linear, and superposition can be applied.

The answer is (A).

2. Resistance is a function of the resistivity, ρ, the length l, and the area, A.

$$R = \frac{\rho l}{A}$$

The length and cross-sectional area are unknown. Therefore, use the following ratio.

$$\frac{R_{50}}{R_{20}} = \frac{\rho_{50}\dfrac{l}{A}}{\rho_{20}\dfrac{l}{A}} = \frac{\rho_{50}}{\rho_{20}}$$

Therefore,

$$R_{50} = R_{20}\frac{\rho_{50}}{\rho_{20}}$$

Solving for ρ_{50} gives

$$
\begin{aligned}
\rho_{50} &= \rho_{20}\Big(1 + \alpha_{20}(T - 20^\circ\text{C})\Big) \\
&= (1.8 \times 10^{-8}\ \Omega\cdot\text{m}) \\
&\quad \times \Big(1 + (3.9 \times 10^{-3}\ ^\circ\text{C}^{-1})(50^\circ\text{C} - 20^\circ\text{C})\Big) \\
&= 20.11 \times 10^{-9}\ \Omega\cdot\text{m}
\end{aligned}
$$

Substituting the given and calculated information gives

$$
\begin{aligned}
R_{50} &= R_{20}\frac{\rho_{50}}{\rho_{20}} = (5\ \Omega)\left(\frac{20.11 \times 10^{-9}\ \Omega\cdot\text{m}}{1.8 \times 10^{-8}\ \Omega\cdot\text{m}}\right) \\
&= \boxed{5.6\ \Omega}
\end{aligned}
$$

3. The capacitance is a function of the permittivity $\epsilon_r\epsilon_0$, the area of the plates, A, and the distance, r, between them.

$$
\begin{aligned}
C &= \frac{\epsilon A}{r} = \frac{\epsilon_r\epsilon_0 A}{r} \\
&= \frac{(3.4)\left(8.854 \times 10^{-12}\ \dfrac{\text{C}^2}{\text{N}\cdot\text{m}^2}\right)(0.04\ \text{m})^2}{0.001\ \text{m}} \\
&= \boxed{4.817 \times 10^{-11}\ \text{F}}
\end{aligned}
$$

4. The charge on the plates is

$$Q = CV_{\text{plates}} = (4.817 \times 10^{-11} \text{ F})(200 \text{ V})$$
$$= \boxed{9.63 \times 10^{-9} \text{ C}}$$

5. The average energy in the inductor's magnetic field depends on the self-inductance, L, and the effective current, I.

$$U = \tfrac{1}{2}LI^2 = \left(\tfrac{1}{2}\right)(8 \times 10^{-3} \text{ H})(15 \text{ A})^2$$
$$= 0.90 \text{ J}$$

The energy in terms of the magnetic flux, Ψ, is

$$U = \tfrac{1}{2}\left(\frac{\Psi^2}{L}\right)$$

$$\Psi = \sqrt{2UL} = \sqrt{(2)(0.90 \text{ J})(8 \times 10^{-3} \text{ H})}$$

$$= \boxed{0.120 \text{ Wb}}$$

6. Resistors in parallel with voltage sources, or in series with current sources, do not affect the remainder of the circuit. Thus, R_1 and R_4 are redundant.

The answer is (D).

7. The circuit can be represented as

From the loop-current method,

$$V_l = R_l I_l + R_c(I_l + I_r)$$
$$V_r = R_r I_r + R_c(I_l + I_r)$$
$$I_l + I_r = \frac{R_r V_l + R_l V_r}{R_r R_l + R_r R_c + R_l R_c}$$
$$= \frac{(4 \text{ }\Omega)(32 \text{ V}) + (2 \text{ }\Omega)(20 \text{ V})}{(4 \text{ }\Omega)(2 \text{ }\Omega) + (4 \text{ }\Omega)(8 \text{ }\Omega) + (2 \text{ }\Omega)(8 \text{ }\Omega)}$$
$$= \boxed{3 \text{ A} \quad [\text{down}]}$$

If the first two equations are solved simultaneously,

$$I_l = 4 \text{ A}$$
$$I_r = -1 \text{ A}$$

8. The circuit can be represented as

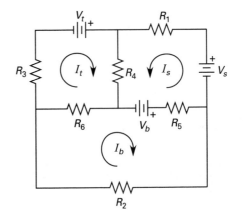

Using the loop-current method,

$$V_t = R_4(I_t + I_s) + R_6(I_t - I_b) + R_3 I_t$$
$$V_s + V_b = R_1 I_s + R_4(I_t + I_s) + R_5(I_s + I_b)$$
$$V_b + R_6(I_t - I_b) = R_5(I_b + I_s) + R_2 I_b$$

After rearranging, substituting in the constants, and evaluating the coefficients, the previous three equations become

$$24 \text{ V} = (12 \text{ }\Omega)I_t + (6 \text{ }\Omega)I_s - (2 \text{ }\Omega)I_b$$
$$14 \text{ V} = (6 \text{ }\Omega)I_t + (10 \text{ }\Omega)I_s + (1 \text{ }\Omega)I_b$$
$$12 \text{ V} = -(2 \text{ }\Omega)I_t + (1 \text{ }\Omega)I_s + (8 \text{ }\Omega)I_b$$

Solving simultaneously,

$$I_t = 2.53 \text{ A}$$
$$I_s = -0.336 \text{ A}$$
$$I_t + I_s = \boxed{2.194 \text{ A} \quad [\text{down}]}$$

9. (a) Before the switch is moved to position B,

$$I(t) = \left(\frac{V_{100}}{R}\right)(1 - e^{-Rt/L})$$

$$I(5 \times 10^{-4}) = \left(\frac{100 \text{ V}}{80 \text{ }\Omega}\right)\left(1 - e^{-(80 \text{ }\Omega)(5 \times 10^{-4} \text{ s})/0.15 \text{ H}}\right)$$

$$= \boxed{0.2926 \text{ A} \quad [\text{clockwise}]}$$

(b) Move the switch to position B. Solve for currents I_1 and I_2, whose sum is I_N, the current flowing in the circuit.

$$I_1(t) = I_0 e^{-Rt/L} \quad [\text{clockwise}]$$

$$I_2(t) = \left(\frac{V_{50}}{R}\right)(1 - e^{-Rt/L}) \quad [\text{clockwise}]$$

$$I_N(t) = I_1(t) + I_2(t)$$

$$= I_0 e^{-Rt/L} + \left(\frac{V_{50}}{R}\right)(1 - e^{-Rt/L})$$

$$= \frac{50 \text{ V}}{80 \text{ }\Omega} + \left(0.2926 \text{ A} - \frac{50 \text{ V}}{80 \text{ }\Omega}\right)e^{-(80 \text{ }\Omega)t/0.15 \text{ H}}$$

$$= \boxed{0.625 - 0.332e^{-(533s^{-1})t} \text{ A} \quad [\text{clockwise}]}$$

10. Consider the circuit shown.

Because all the impedances are resistors, the phasor notation is dropped for convenience. From the deriving equations, the impedance is

$$z_{11} = \left.\frac{V_1}{I_1}\right|_{I_2 = 0 \text{ A}}$$

The circuit, redrawn with $I_2 = 0$ A, is

The equivalent impedance looking to the right, into the input terminals, is

$$z_e = R_e = 5 \text{ }\Omega + \frac{(4 \text{ }\Omega)(2 \text{ }\Omega + 1 \text{ }\Omega + 1 \text{ }\Omega)}{4 \text{ }\Omega + 2 \text{ }\Omega + 1 \text{ }\Omega + 1 \text{ }\Omega}$$

$$= 7 \text{ }\Omega$$

Redrawing the circuit again gives

Using Ohm's law gives

$$V_1 = I_1(7 \text{ }\Omega)$$

$$\frac{V_1}{I_1} = 7 \text{ }\Omega$$

Therefore,

$$z_{11} = \boxed{7 \text{ }\Omega}$$

From the deriving equations, the impedance is

$$z_{21} = \left.\frac{V_2}{I_1}\right|_{I_2 = 0 \text{ A}}$$

The redrawn circuit is identical to that for z_{11}, since $I_2 = 0$ A.

Using the concept of a current divider at node A gives the following for the current through the branch containing V_2.

$$I_{V_2} = I_1\left(\frac{4 \text{ }\Omega}{4\Omega + (2 \text{ }\Omega + 1 \text{ }\Omega + 1 \text{ }\Omega)}\right) = \tfrac{1}{2}I_1$$

Apply Ohm's law across the 1 Ω resistor associated with V_2, using the calculated current.

$$V_2 = \tfrac{1}{2}I_1(1 \text{ }\Omega) = \tfrac{1}{2}I_1$$

$$\frac{V_2}{I_1} = \tfrac{1}{2}\Omega$$

Therefore,

$$z_{21} = \left.\frac{V_2}{I_1}\right|_{I_2=0 \text{ A}} = \tfrac{1}{2}\ \Omega$$

From the deriving equations,

$$z_{12} = \left.\frac{V_1}{I_2}\right|_{I_1=0 \text{ A}}$$

The circuit, redrawn with $I_1 = 0$ A, is

Using a current divider at node B gives the following for the branch current flowing through the 4 Ω resistor associated with V_1.

$$I_{V_1} = I_2\left(\frac{1\ \Omega}{1\ \Omega + 2\ \Omega + 4\ \Omega + 1\ \Omega}\right) = \tfrac{1}{8}I_2$$

Applying Ohm's law for the 4 Ω resistor gives

$$V_1 = \tfrac{1}{8}I_2(4\ \Omega) = \tfrac{1}{2}I_2$$

$$\frac{V_1}{I_2} = \tfrac{1}{2}\ \Omega$$

Therefore,

$$z_{12} = \left.\frac{V_1}{I_2}\right|_{I_1=0 \text{ A}} = \tfrac{1}{2}\ \Omega$$

From the deriving equations,

$$z_{22} = \left.\frac{V_2}{I_2}\right|_{I_1=0 \text{ A}}$$

The redrawn circuit is identical to that for z_{12}, since $I_1 = 0$ A.

The equivalent impedance looking to the left, into the output terminals, is

$$z_e = R_e = \frac{(1\ \Omega)(2\ \Omega + 4\ \Omega + 1\ \Omega)}{1\ \Omega + (2\ \Omega + 4\ \Omega + 1\ \Omega)} = \tfrac{7}{8}\ \Omega$$

Redrawing the circuit again gives

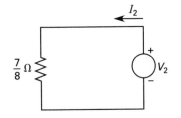

Using Ohm's law gives

$$V_2 = I_2\left(\tfrac{7}{8}\ \Omega\right)$$

$$\frac{V_2}{I_2} = \tfrac{7}{8}\ \Omega$$

Therefore,

$$z_{22} = \tfrac{7}{8}\ \Omega$$

Circuit Theory

30 Transient Analysis

PRACTICE PROBLEMS

1. Consider the circuit shown. What is the time constant?

- (A) 1.0 μs
- (B) 36 μs
- (C) 0.30 s
- (D) 3.0 s

2. With an initially uncharged capacitor in the circuit in Prob. 1, what is the initial current?

- (A) 0.0 A
- (B) 4.0 μA
- (C) 6.0 μA
- (D) 120 A

3. How long does it take the circuit shown in Prob. 1 to reach steady-state response?

4. What is the rise time of the circuit in Prob. 1?

5. Consider the circuit shown. Write the equations for loops 1 and 2.

6. A series RC circuit contains a 4 Ω resistor and an uncharged 300 μF capacitor. The switch is closed at $t=0$ s, connecting the circuit across a 500 V source. For the circuit shown, determine the current as a function of time.

7. The switch in the circuit shown is closed at $t=0$ s and opened again at $t=0.05$ s. How much energy is stored in the initially uncharged capacitor?

8. Find the current through the 80 Ω resistor (a) immediately after the switch is closed and (b) 2 s after the switch is closed. The switch has been open for a long time.

9. For the circuit shown, determine the steady-state inductance current, i_L, (a) with the switch closed and (b) with the switch open.

10. For the circuit shown, determine the steady-state capacitance voltage, V_C, (a) with the switch closed and (b) with the switch open.

11. For the circuit shown, determine the steady-state inductance current, i_L, (a) with the switch closed and (b) with the switch open.

12. The circuit in Prob. 10 has had the switch closed for a long time. At $t=0$ s, the switch opens. Determine the capacitance voltage, $v_C(t)$, for $t > 0$ s, for a capacitance of 1×10^{-6} F.

13. The circuit in Prob. 10 has had the switch open for a long time. At $t=0$ s, the switch closes. Determine the capacitance voltage, $v_C(t)$, for $t > 0$ s.

14. A system is described by the equation

$$12 \sin 2t = \frac{d^2 v}{dt^2} + 5 \frac{dv}{dt} + 4v$$

The initial conditions on the voltage are

$$v(0) = -5 \text{ V}$$

$$\frac{dv(0)}{dt} = 2 \text{ V/s}$$

Determine $v(t)$ for $t > 0$ s.

15. For the circuit shown, the current source is

$$i_s(t) = 0.01 \cos 500t$$

Find the transfer function.

$$G(s) = \frac{V_C(s)}{I_s(s)}$$

16. For a series RLC circuit of 10 Ω, 0.5 H, and 100 μF, determine the (a) resonant frequency, (b) quality factor, and (c) bandwidth.

17. For a parallel RLC circuit of 10 Ω, 0.5 H, and 100 μF, determine the (a) resonant frequency, (b) quality factor, and (c) bandwidth.

18. The switch of the circuit shown has been open for a long time and is closed at $t=0$ s. Determine the currents flowing in the two capacitances at the instant the switch closes. Give the magnitudes and directions.

19. A system is described by the following differential equation.

$$20 \sin 4t = \frac{d^2 i}{dt^2} + 4 \frac{di}{dt} + 4i$$

The initial conditions are

$$i(0) = 0 \text{ A}$$

$$\frac{di(0)}{dt} = 4$$

Determine $i(t)$ for $t > 0$ s.

20. The circuit shown has the 1 μF capacitance charged to 100 V and the 2 μF capacitance uncharged before the switch is closed. Determine the (a) energy stored in each capacitance at the instant the switch closes, (b) energy stored in each capacitance a long time after the switch closes, and (c) energy dissipated in the resistance.

SOLUTIONS

1.

The time constant is

$$\tau = RC = (3 \times 10^6 \ \Omega)(0.1 \times 10^{-6} \ \text{F})$$
$$= \boxed{0.3 \ \text{s} \quad (0.30 \ \text{s})}$$

The answer is (C).

2. The capacitor initially acts as a short. From Ohm's law,

$$V = IR$$
$$I = \frac{V}{R} = \frac{12 \ \text{V}}{3 \times 10^6 \ \Omega}$$
$$= \boxed{4.0 \times 10^{-6} \ \text{A} \quad (4.0 \ \mu\text{A})}$$

The answer is (B).

3. Steady-state response is reached in approximately 5 time constants. Thus,

$$t = 5\tau$$

The time constant is $RC = 0.3$ s. Substituting,

$$t = (5)(0.3 \ \text{s}) = \boxed{1.5 \ \text{s}}$$

4. The rise time is most often determined by measurements in the lab. Nevertheless, the rise time is

$$t_r = 2.2\tau$$

The time constant was found in Prob. 1. Substituting gives

$$t_r = (2.2)(0.3 \ \text{s}) = \boxed{0.660 \ \text{s}}$$

Circuit Theory

5.

Kirchoff's voltage law (KVL) for loop 1 gives

$$v_s - L_1\left(\frac{di_1}{dt}\right)$$
$$-(i_1 - i_2)R_1 = 0$$
$$v_s = \boxed{L_1\left(\frac{di_1}{dt}\right) + (i_1 - i_2)R_1}$$

KVL for loop 2 gives

$$-(i_2 - i_1)R_1 - L_2\left(\frac{di_2}{dt}\right) - i_2 R_2 = 0$$
$$-i_2 R_1 + i_1 R_1 - L_2\left(\frac{di_2}{dt}\right) - i_2 R_2 = 0$$
$$\boxed{L_2\left(\frac{di_2}{dt}\right) + R_1(i_2 - i_1) + i_2 R_2 = 0}$$

6. From Table 30.1, the current response for a series RC circuit is

$$i(t) = \left(\frac{V_{\text{bat}} - V_0}{R}\right)e^{-t/\tau} = \left(\frac{500\text{ V} - 0\text{ V}}{4\text{ }\Omega}\right)e^{-t/(1.2\times10^{-3})}$$
$$= \boxed{125e^{-833t}\text{ A}}$$

7.
$$Q(t) = CV_C(t) = CV(1 - e^{-t/RC})$$
$$E_C(t) = \tfrac{1}{2}CV_C^2(t) = \tfrac{1}{2}CV^2(1 - e^{-t/RC})^2$$
$$E_C(0.05) = \left(\tfrac{1}{2}\right)(100\times10^{-6}\text{ F})(100\text{ V})^2$$
$$\times\left(1 - e^{-0.05\text{ s}/((10^3\text{ }\Omega)(100\times10^{-6}\text{ F}))}\right)^2$$
$$= \boxed{0.0774\text{ J}}$$

8. (a) Since the current in the inductor cannot change instantaneously,

$$I(0) = \boxed{0\text{ A}}$$

(b) $\quad I(t) = \left(\frac{V}{R}\right)(1 - e^{-Rt/L})$

$$I(2) = \left(\frac{100\text{ V}}{80\text{ }\Omega}\right)\left(1 - e^{-(80\text{ }\Omega)(2\text{ s})/7\text{ H}}\right)$$
$$= \boxed{1.25\text{ A}\quad[\text{down}]}$$

9. (a) Use KVL around the outer loop. With the switch closed, KVL gives

$$15\text{ V} = 4i_L + v_L$$
$$v_L = L\left(\frac{di_L}{dt}\right) = 0\quad[\text{in steady state, ss}]$$
$$i_{L_{ss}} = i_{ss} = \frac{15\text{ A}}{4\text{ A}}$$

The current through the 8 Ω resistor is $v_L/8 = 0$ V; therefore, in the steady-state condition, all of i_{ss} flows through the inductance.

(b) With the switch open, the inductor dissipates its energy in the 8 Ω resistor.

$$i_{L_{ss}} = \boxed{i_{ss} = 0}$$

10. (a) With the switch closed,

$$i_C = C\left(\frac{dv_C}{dt}\right)$$

In steady-state condition,

$$\frac{dv_C}{dt} = 0$$

Therefore, $i_C \to 0$. By voltage division,

$$v_C = \left(\frac{8\text{ }\Omega}{12\text{ }\Omega}\right)(15\text{ V}) = \boxed{10\text{ V}}$$

With the switch open, the capacitor dissipates its energy in the 8 Ω resistor.

(b)

$$\boxed{v_C = 0}$$

11. (a) A Thevenin equivalent circuit, seen from the inductance, is obtained.

$$v_{oc} = v_{\text{Th}} = \left(\frac{8\text{ }\Omega}{12\text{ }\Omega}\right)v_s = 10\cos(10t + 20°)\text{ V}$$
$$i_{sc} = \frac{15\cos(10t + 20°)\text{ V}}{4\text{ }\Omega} = 3.75\cos(10t + 20°)\text{ A}$$
$$R_{\text{Th}} = \frac{v_{oc}}{i_{sc}} = \frac{10\text{ V}}{3.75\text{ A}} = 8/3\text{ }\Omega$$

Taking $\cos(10t + 20°)$ as the phasor reference and using the maximum value, for convenience, instead of the rms value, gives

$$V_{\text{Th}} = 10 \text{ V} \angle 0°$$

$$Z_L = j\omega L = \left(j10 \ \frac{\text{rad}}{\text{s}}\right)(0.1 \text{ H}) = j1 \ \Omega$$

$$i_L = \frac{V_{\text{Th}}}{R_{\text{Th}} + j1 \ \Omega}$$

$$= \left(\frac{10 \text{ V}}{\frac{8}{3} \ \Omega + j1 \ \Omega}\right)\left(\frac{\frac{8}{3} \ \Omega - j1 \ \Omega}{\frac{8}{3} \ \Omega - j1 \ \Omega}\right)$$

$$= \left(\tfrac{90}{73} \text{ A}\right)\left(\tfrac{8}{3} - j1\right) = \boxed{3.51 \text{ A} \angle -20.6°}$$

As a time function,

$$i_L(\infty) = 3.51 \cos(10t + 20° - 20.6°)$$

$$= 3.51 \cos(10t - 0.6°) \text{ A}$$

(b) With the switch open,

$$i_L(\infty) = \boxed{0}$$

12. With the switch closed for a long time, the capacitance voltage is 10 V (see Prob. 10). The initial voltage when the switch opens is therefore 10 V.

With the switch open, KVL is

$$v_C = 8i$$

$$i_C = 10^{-6}\left(\frac{dv_C}{dt}\right)$$

$$i_C = -i$$

$$v_C + (8 \times 10^{-6})\left(\frac{dv_C}{dt}\right) = 0$$

$$\frac{dv_C}{dt} + (1.25 \times 10^5)v_C = 0$$

Any appropriate mathematical method may now be used to solve the differential equation.

Let $\alpha = 1.25 \times 10^5$. The solution is of the form

$$v_C = V(t_1)e^{-\alpha(t - t_1)}$$

Taking $t_1 = 0$ s, $v_C(0) = 10$ V.

$$\frac{dv_C}{dt} = -\alpha v_C$$

$$\alpha = 1.25 \times 10^5$$

$$v_C(t) = \boxed{10e^{-(1.25 \times 10^5)t} \text{ V}}$$

13. The initial capacitance voltage is zero. Kirchhoff's current law (KCL) at the upper node, with the lower node as a reference, is

$$\frac{15 \text{ V} - v_C}{4 \ \Omega} = \frac{v_C}{8 \ \Omega} + (10^{-6} \text{ F})\left(\frac{dv_C}{dt}\right)$$

This is manipulated to

$$\left(\tfrac{8}{3} \times 10^{-6}\right)\left(\frac{dv_C}{dt}\right) + v_C = 10 \text{ V}$$

$$V_{\text{ss}} = 10 \text{ V}$$

$$\tau = \tfrac{8}{3} \times 10^{-6} \text{ s} \quad (2.67 \times 10^{-6} \text{ s})$$

$$V_0 = 0$$

$$v_C(t) = V_{\text{ss}} + (V_0 - V_{\text{ss}})e^{-t/\tau}$$

$$= \boxed{(10)\left(1 - e^{-(3.75 \times 10^5)t}\right)u(t) \text{ V}}$$

14. The homogeneous differential equation is

$$\frac{d^2 v}{dt^2} + 5\frac{dv}{dt} + 4v = 0$$

This has solutions e^{st},

$$\frac{d^2 v}{dt^2} \to s^2 V$$

$$\frac{dv}{dt} \to sV$$

$$s^2 + 5s + 4 = 0$$

$$(s + 4)(s + 1) = 0$$

Transient solutions are of the form

$$Ae^{-4t} + Be^{-t}$$

The steady-state sinusoidal solution must be found before A and B can be evaluated.

The phasor solution is

$$12 \sin 2t \Leftrightarrow 12 \text{ V} \angle 0°$$

$$\omega = 2$$

$$\frac{d}{dt} \Rightarrow j\omega = j2$$

$$\frac{d^2}{dt^2} \Rightarrow -\omega^2 = -4$$

Circuit Theory

The phasor equation is

$$12 \text{ V} = -4V_{ss} + (5)(j2)V_{ss} + 4V_{ss}$$

$$= j10V_{ss}$$

$$V_{ss} = \frac{1.2}{j} = 1.2 \text{ V} \angle -90°$$

$$v_{ss} = 1.2 \text{ V} \sin(2t - 90°)$$

$$= -1.2 \text{ V} \cos 2t$$

$$v(t) = Ae^{-4t} + Be^{-t} - 1.2 \cos 2t$$

$$v(0) = A + B - 1.2 \text{ V} = -5 \text{ V}$$

$$A + B = -3.8 \text{ V}$$

$$\frac{dv}{dt} = -4Ae^{-4t} - Be^{-t} + 2.4 \sin 2t$$

$$\frac{dv}{dt}(0) = -4A - B = 2$$

$$A = 0.6 \text{ V}$$

$$B = -4.4 \text{ V}$$

$$v(t) = \boxed{(0.6e^{-4t} - 4.4e^{-t} - 1.2\cos 2t)u(t) \text{ V}}$$

15. The impedance of the inductance branch is

$$Z_{LR} = 10 + 0.25s \text{ } \Omega$$

KCL applied to the upper node is

$$I_s = \frac{V_C}{30 \text{ } \Omega} + (0.001 \text{ F})sV_C + \frac{V_C}{10 \text{ } \Omega + (0.25 \text{ H})s}$$

Solving,

$$G = \frac{V_C}{I_s} = \left(\frac{3 \times 10^4 + 750s}{0.75s^2 + 55s + 4000}\right) \Omega$$

$$= \boxed{(1000)\left(\frac{s + 40}{s^2 + 73.33s + 5333}\right) \Omega}$$

16. (a) $\quad Z = 10 \text{ } \Omega + 0.5s \text{ } \Omega + \dfrac{10^4}{s} \text{ } \Omega$

$$= \left(\frac{1}{2}\right)\left(\frac{s^2 + 20s + 2 \times 10^4}{s}\right) \Omega$$

$$= \left(\frac{1}{2s}\right)\left(s^2 + \frac{\omega_0}{Q}s + \omega_0^2\right) \Omega$$

The resonant frequency is

$$\omega_0 = \boxed{\sqrt{2} \times 10^2 \text{ rad/s}}$$

(b) The quality factor is

$$Q = \frac{\omega_0}{20} = \boxed{5\sqrt{2}}$$

(c) The bandwidth is

$$\text{BW} = \frac{\omega_0}{Q} = \boxed{20 \text{ rad/s}}$$

17. (a) $\quad Y = 0.1 + \dfrac{1}{0.5s} + 10^{-4}s$

$$= \left(\frac{10^{-4}}{s}\right)(s^2 + 10^3 s + 2 \times 10^4)$$

$$= \left(\frac{10^{-4}}{s}\right)\left(s^2 + \frac{\omega_0}{Q}s + \omega_0^2\right)$$

The resonant frequency is

$$\omega_0 = \sqrt{2 \times 10^4} = \boxed{100\sqrt{2} \text{ rad/s}}$$

(b) The quality factor is

$$Q = \frac{\omega_0}{\text{BW}} = \frac{\sqrt{2}}{10} = \boxed{0.1\sqrt{2}}$$

(c) The bandwidth is

$$\text{BW} = \frac{\omega_0}{Q} = \boxed{10^3 \text{ rad/s}}$$

18. With the switch open for a long time, the upper-right capacitor is charged to 10 V and the lower (center) capacitor is discharged to 0 V. At the instant of switching, the capacitors can be modeled as ideal voltage sources with their initial voltages. The resulting equivalent circuit is shown.

KVL on the left loop is

$$10 \text{ V} - (50 \text{ } \Omega)i_s + 0.0 = 0$$

$$i_s = 1/5 \text{ A}$$

KVL on the right loop is

$$0.0 + 10 \text{ V} - (100 \text{ } \Omega)i_2 = 0$$

$$i_2 = 1/10 \text{ A}$$

The current in the right (upper) capacitor is i_2, with 0.1 A flowing left.

As the voltage across the 1 kΩ resistor is zero, $i_B = 0$ A. By KCL, the other capacitor current is

$$i_A = i_s + i_2 = \tfrac{1}{5}\,\text{A} + \tfrac{1}{10}\,\text{A} = \boxed{0.3\ \text{A}}$$

19. The homogeneous equation is

$$s^2 + 4s + 4 = 0$$

The roots are $(s+2)^2$. The homogeneous solution is

$$Ae^{-2t} + Bte^{-2t}$$

The steady-state solution is

$$s \to j\omega$$
$$\omega = 4$$
$$\frac{d}{dt} \to j\omega$$
$$\frac{d^2}{dt^2} \to -\omega^2 = -16$$

The sine reference is used.

$$20\ \text{V} = (-16\ \Omega + 16j\ \Omega + 4\ \Omega)I_{\text{ss}}$$
$$I_{\text{ss}} = \frac{20}{-12 + j16} = -0.6 + j(-0.8)\ \text{A}$$
$$i_{\text{ss}} = -0.6 \sin 4t - 0.8 \sin(4t + 90°)\ \text{A}$$
$$= -0.6 \sin 4t - 0.8 \cos 4t\ \text{A}$$
$$i = i_{\text{ss}} + Ae^{-2t} + Bte^{-2t}$$
$$i(0) = -0.8\ \text{A} + A = 0$$
$$A = 0.8\ \text{A}$$
$$\frac{di}{dt} = -2.4 \cos 4t + 3.2 \sin 4t$$
$$\qquad - 2Ae^{-2t} + Be^{-2t} - 2Bte^{-2t}$$
$$\frac{di}{dt}(0) = -2.4 - 2A + B = 4\ \text{A}$$
$$B = 2.4 + 2A + 4 = 8\ \text{A}$$
$$i(t) = 0.8e^{-2t} + 8te^{-2t} - 0.6 \sin 4t - 0.8 \cos 4t\ \text{A}$$
$$= \boxed{0.8e^{-2t} + 8te^{-2t} - \sin(4t + 53.1°)\ \text{A}}$$

20. The initial charge on the 1 μF capacitor is

$$Q_{1,\text{initial}} = CV_0$$
$$= (10^{-6}\ \text{F})(100\ \text{V})$$
$$= 10^{-4}\ \text{C}$$

The initial energy, $W_{1,\text{initial}}$, is

(a)
$$W_{1,\text{initial}} = \tfrac{1}{2}C_1 V_0^2$$
$$= \left(\tfrac{1}{2}\right)(10^{-6}\ \text{F})(100\ \text{V})^2$$
$$= 0.005\ \text{J} \quad (5\ \text{mJ})$$

In the final condition, the voltage across the two capacitors is the same.

$$W_{2,\text{initial}} = 0$$
$$V_f = \frac{Q_1}{C_1} = \frac{Q_2}{C_2}$$
$$Q_2 = 2Q_1$$

As $Q_1 + Q_2 = Q_0$,
$$Q_1 = \tfrac{1}{3} \times 10^{-4}\ \text{C}$$
$$Q_2 = \tfrac{2}{3} \times 10^{-4}\ \text{C}$$

$$W_{\text{final}} = \tfrac{1}{2}\left(\frac{Q_1^2}{C_1}\right) + \tfrac{1}{2}\left(\frac{Q_2^2}{C_2}\right)$$
$$= \left(\tfrac{1}{2}\right)\left(\frac{\left(\tfrac{1}{3} \times 10^{-4}\ \text{C}\right)^2}{1 \times 10^{-6}\ \text{F}}\right) + \left(\tfrac{1}{2}\right)\left(\frac{\left(\tfrac{2}{3} \times 10^{-4}\ \text{C}\right)^2}{2 \times 10^{-6}\ \text{F}}\right)$$
$$= 0.001667\ \text{J} \quad (1.667\ \text{mJ})$$
$$W_{\text{loss}} = W_{\text{initial}} - W_{\text{final}}$$
$$= 5\ \text{mJ} - 1.667\ \text{mJ}$$
$$= \boxed{3.333\ \text{mJ}}$$

(b) $\quad W_{\text{final}} = \tfrac{3}{18} \times 10^{-2}\ \text{J} = \boxed{0.00167\ \text{J}}$

(c) $\quad W_{\text{loss}} = 0.005\ \text{J} - 0.00167\ \text{J} = \boxed{0.00333\ \text{J}}$

Circuit Theory

31 Time Response

PRACTICE PROBLEMS

1. Which of the following is in the form of the solution to a first-order circuit?

(A) $\kappa + Ae^{-t/\tau}$

(B) $\kappa + Ae^{s_1 t} + Be^{s_2 t}$

(C) $\kappa + Ae^{st} + Bte^{st}$

(D) $\kappa + e^{\alpha t}(A\cos\beta t + B\sin\beta t)$

2. In the circuit shown, how long will it take for the capacitor voltage to reach 98% of its final value, assuming the capacitor is initially uncharged?

3. Determine the time-domain equation for the circuit shown.

4. In the circuit shown in Prob. 3, the inductance is 80 mH and the capacitance is 20 μF. What is the minimum value of the resistance required for the circuit to be overdamped?

(A) 25 Ω

(B) 50 Ω

(C) 100 Ω

(D) 150 Ω

5. Complete the following table for a sine reference.

function	$A_p\angle\phi$	s
5		
$5e^{-200t}$		
$10\sin 377t$		
$10\sin(500t+30°)$		
$20\cos(1000t+60°)$		
$20e^{-200t}\sin 1000t$		

SOLUTIONS

1. The form of a first-order circuit solution is

$$x_t = \boxed{\kappa + Ae^{-t/\tau}}$$

The answer is (A).

2. The time constant, τ, is

$$\tau = RC = (4\ \Omega)(50 \times 10^{-6}\ \text{F})$$
$$= 200 \times 10^{-6}\ \text{s} \quad (200\ \mu\text{s})$$

The build-up factor after a step change takes the form

$$1 - e^{-t/\tau}$$

Therefore, the time to 98% of the final value, regardless of the actual value, is

$$0.98 = 1 - e^{-t/\tau}$$

For the given circuit, $\tau = 200\ \mu$s. Substituting and solving for the time gives

$$0.98 = 1 - e^{-t/200 \times 10^{-6}\ \text{s}}$$
$$-0.02 = -e^{-t/200 \times 10^{-6}\ \text{s}}$$
$$\ln 0.02 = -\frac{t}{200 \times 10^{-6}\ \text{s}}$$
$$t = -\ln 0.02 (200 \times 10^{-6}\ \text{s})$$
$$= \boxed{782.4 \times 10^{-6}\ \text{s} \quad (782\ \mu\text{s} \quad [\approx 4\tau])}$$

3.

Using Kirchoff's voltage law (KVL) around the circuit gives

$$v_s - iR - v_L - v_C = 0\ \text{V}$$
$$v_s = iR + v_L + v_C$$
$$= iR + L\left(\frac{di}{dt}\right) + \frac{1}{C}\int i\,dt$$

Differentiating gives

$$\frac{dv_s}{dt} = \left(\frac{di}{dt}\right)R + L\left(\frac{d^2i}{dt^2}\right) + \frac{i}{C}$$
$$= \boxed{L\left(\frac{d^2i}{dt^2}\right) + R\left(\frac{di}{dt}\right) + \frac{1}{C}i}$$

4. The overdamped condition is determined by the coefficients of the differential equation. Specifically, for the overdamped condition, $b^2 > 4ac$.

From the solution of Prob. 3,

$$R^2 > 4L\left(\frac{1}{C}\right)$$

$$R < \sqrt{(4)(80 \times 10^{-3}\ \text{H})\left(\frac{1}{20 \times 10^{-6}\ \text{F}}\right)} = \boxed{126.5\ \Omega}$$

The answer is (D).

5.

function	$A_p \angle \phi$	s
5	$5\angle 0°$	$0 + j0$
$5e^{-200t}$	$5\angle 0°$	$-200 + j0$ Np/s
$10\sin 377t$	$10\angle 0°$	$0 \pm j377$ rad/s
$10\sin(500t + 30°)$	$10\angle 30°$	$0 \pm j500$ rad/s
$20\cos(1000t + 60°)$	$20\angle 150°$	$0 \pm j1000$ rad/s
$20e^{-200t}\sin 1000t$	$20\angle 0°$	$-200 \pm j1000$ s^{-1}

32 Frequency Response

PRACTICE PROBLEMS

1. For the circuit shown, determine the transfer function with $V(s)$ as the input and $I(s)$ as the output.

2. Plot the zeros and poles on the s domain for the circuit in Prob. 1.

3. Plot the Bode magnitude plot for the circuit in Prob. 1.

4. For the circuit shown, plot the idealized Bode plot of magnitude and phase.

5. In the Bode plot for the circuit in Prob. 4, what type of filter is represented?

(A) low-pass filter

(B) high-pass filter

(C) bandpass filter

(D) notch filter

6. For the circuit shown, determine the (a) voltage transfer function, (b) frequency response, and (c) unit step voltage response.

7. For the circuit shown, determine the (a) voltage transfer function, (b) frequency response, and (c) unit step voltage response.

SOLUTIONS

1. The transfer function is

$$T_{\mathrm{net}}(s) = \frac{I(s)}{V(s)} = \frac{1}{Z(s)}$$

First, transfer the circuit into the s domain.

The impedance for this series circuit is found in terms of the complex frequency, s.

$$Z(s) = R + Z_C + Z_L = 5 + \frac{6}{s} + 1s$$

$$= \frac{5s + 6 + s^2}{s} = \frac{s^2 + 5s + 6}{s}$$

The transfer function is

$$\boxed{T_{\mathrm{net}} = \frac{1}{Z(s)} = \frac{s}{s^2 + 5s + 6}}$$

2. Factor the transfer function into standard form.

$$T_{\mathrm{net}}(s) = \frac{s}{s^2 + 5s + 6}$$

$$= \frac{s}{(s+2)(s+3)}$$

One zero at the origin is given by the numerator term. Two poles, one at -2 and the other at -3, exist. Plotting gives

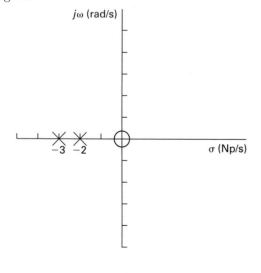

3. The transfer function was determined in Prob. 2 to be

$$T_{\mathrm{net}}(s) = \frac{s}{(s+2)(s+3)}$$

The zero break frequency is $\omega = 1$. The break frequency for the poles is $\omega = 2$ and $\omega = 3$. These points are plotted on the following diagram.

For the zero, a $+20$ dB/decade line through the origin is plotted. For the poles, the low-frequency asymptotes are plotted on the frequency axis extending to lower frequencies from $\omega = 2$ and $\omega = 3$. The high-frequency asymptotes break down at -20 dB/decade from these points. Summing the asymptotic lines gives the net Bode magnitude plot as shown.

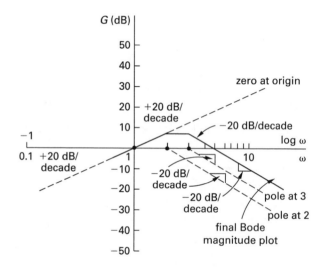

The value of the constant, K, is determined from

$$T_{\mathrm{net}}(j\omega) = K\left(\frac{j\omega}{\left(1 + \dfrac{j\omega}{2}\right)\left(1 + \dfrac{j\omega}{3}\right)}\right)$$

$$K = A\left(\frac{\displaystyle\prod_1^n z_n}{\displaystyle\prod_1^d p_d}\right)$$

The transfer $T_{\mathrm{net}}(s)$ gives the value of A as 1. For a zero at the origin in the s domain, the break frequency is ω equals 1. Therefore, z_n equals 1.

$$K = (1)\left(\frac{1}{(2)(3)}\right) = 1/6$$

This makes $20\log K$ equal to -15.563 dB. Either the Bode plot should be shifted downward by -15.563 or the zero dB point should be changed to -15.563 to obtain the final solution.

Circuit Theory

4. Because the calculations will take place in the s domain, transform the circuit.

The transfer function, using the voltage divider concept, is as shown.

$$T_{\text{net}}(s) = \frac{V_2(s)}{V_1(s)} = \frac{\dfrac{10}{s}}{\dfrac{10}{s} + 10} = \frac{\dfrac{10}{s}}{\dfrac{10 + s10}{s}}$$

$$= \frac{10}{10 + s10}$$

$$= \frac{1}{s + 1}$$

The break frequency, or pole, occurs at $\omega = 1$. At this point the phase is $-45°$. Using the standard method, the resulting plot is as shown.

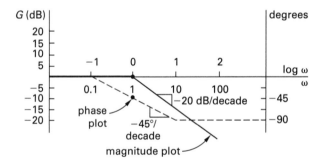

5. The Bode plot of Prob. 4 indicates that low frequencies are passed with a gain of 0 dB. At high frequencies, the output is attenuated and the gain steadily drops. The circuit is called a low-pass filter.

The answer is (A).

6. (a) By voltage division,

$$\left(\frac{V_{\text{out}}}{V_{\text{in}}}\right)s = \frac{0.01s}{0.01s + 50 + \dfrac{10^6}{s}}$$

$$= \boxed{\frac{s^2}{s^2 + 5000s + 10^8}}$$

(b) The denominator needs an underdamped response, so it is put into standard form to take advantage of the Laplace transform pairs table.

$$\left(\frac{V_{\text{out}}}{V_{\text{in}}}\right)s = \frac{s^2}{(s + 2500)^2 + (9682)^2}$$

For the frequency response $s \to j\omega$,

$$\left(\frac{V_{\text{out}}}{V_{\text{in}}}\right)j\omega = \frac{-\omega^2}{10^8 - \omega^2 + j5000\omega}$$

At low frequency,

$$\frac{V_{\text{out}}}{V_{\text{in}}} \to \frac{-\omega^2}{10^8}$$

This has a slope of 40 dB per decade and a phase of $180°$ on a Bode diagram.

At high frequency,

$$\frac{V_{\text{out}}}{V_{\text{in}}} \to 1$$

At $\omega = 10^4$ rad/s,

$$\frac{V_{\text{out}}}{V_{\text{in}}} = 2\angle 90°$$

The frequency response value is $\boxed{6 \text{ dB.}}$

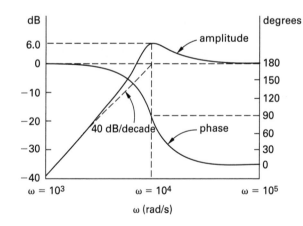

(c) The step response with $V_{\text{in}} = 1/s$ is

$$V_{\text{out}} = \frac{s}{(s + 2500)^2 + (9682)^2}$$

$$= \frac{s + 2500}{(s + 2500)^2 + (9682)^2}$$

$$- \left(\frac{2500}{9682}\right)\left(\frac{9682}{(s + 2500)^2 + (9682)^2}\right)$$

$$v_{\text{out}}(t) = e^{-2500t}(\cos 9682t - 0.258 \sin 9682t) \text{ V}$$

$$= \boxed{1.03 e^{-2500t} \cos(9682t + 14.5°) \text{ V}}$$

Circuit Theory

7. (a) By voltage division,

$$\frac{V_{\text{out}}}{V_{\text{in}}} = \frac{\dfrac{10^6}{s}}{10^{-2}s + 50 + \dfrac{10^6}{s}}$$

$$= \boxed{\frac{10^8}{s^2 + 5000s + 10^8}}$$

$$\left(\frac{V_{\text{out}}}{V_{\text{in}}}\right) j\omega = \frac{10^8}{10^8 - \omega^2 + j5000\omega}$$

(b) At low frequency,

$$\frac{V_{\text{out}}}{V_{\text{in}}} \rightarrow 1\angle 0°$$

At high frequency,

$$\frac{V_{\text{out}}}{V_{\text{in}}} \rightarrow \frac{10^8}{-\omega^2}$$

On the Bode diagram, this has a slope of -40 dB per decade and an angle of $-180°$. At $\omega = 10^4$ rad/s,

$$\frac{V_{\text{out}}}{V_{\text{in}}} = 2\angle -90°$$

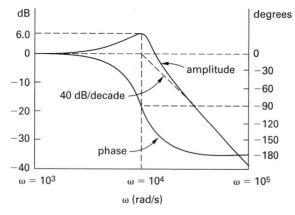

(c) The step response with $V_{\text{in}} = 1/s$ is

$$V_{\text{out}} = \frac{10^8}{s(s^2 + 5000s + 10^8)}$$

This is expanded in partial fractions to

$$V_{\text{out}} = \frac{1}{s} - \frac{s + 5000}{(s + 2500)^2 + (9682)^2}$$

This is recast to fit the Laplace transform pairs table.

$$V_{\text{out}} = \frac{1}{s} - \frac{s + 2500}{(s + 2500)^2 + (9682)^2}$$

$$- \frac{\left(\dfrac{2500}{9682}\right)(9682)}{(s + 2500)^2 + (9682)^2}$$

$$v_{\text{out}}(t) = 1 - e^{-2500t}(\cos 9682t + 0.258 \sin 9682t) \text{ V}$$

$$= \boxed{1 - 1.033 e^{-2500t} \cos(9682t - 14.5°) \text{ V}}$$

33 Generation Systems

PRACTICE PROBLEMS

1. What term describes the process of minimizing irreversibilities by raising the temperature of the water returning to a steam generator?

(A) superheat

(B) reheat

(C) regeneration

(D) both (A) and (C)

2. What fission rate (fissions/s) is required for a nuclear plant to have a 1000 MW thermal output?

3. What type of first stage, or control stage, is normally used on large turbines that are used for electric power generation?

(A) impulse

(B) Rateau

(C) Curtis

(D) Parsons

4. In the load sharing diagram shown, the speed droop for both generators is 0.6 Hz/1000 kW. If generator A is to carry all the load, what no-load frequency setpoint is required on generator A while maintaining the system frequency at 60 Hz?

5. Draw the load sharing diagram for Prob. 4 with generator A carrying all the load. Label all significant points.

6. The field of a DC generator is normally located on the

(A) stator

(B) rotor

(C) armature

(D) round rotor

7. What is the maximum speed for 60 Hz AC generator operation?

8. What power quality term describes a transient of short duration and high magnitude?

(A) overvoltage

(B) harmonic

(C) sag

(D) surge

SOLUTIONS

1. Superheat raises the temperature of the steam supplied to the turbine. Reheat raises the temperature of the steam at the input of the low-temperature stages of a turbine. Regeneration raises the temperature of the water returning to the steam generator.

The answer is (C).

2. The fission rate, fr, is in units of fissions/s. The average energy released per fission is 200 MeV. Unit analysis gives the power output, P_s.

$$P_s = \left(\frac{\text{fissions}}{\text{s}}\right)\left(\frac{\text{energy}}{\text{fission}}\right)$$

$$= \text{fr}\left(200 \ \frac{\text{MeV}}{\text{fission}}\right)$$

$$\text{fr} = \frac{P_s}{200 \ \dfrac{\text{MeV}}{\text{fission}}}$$

$$= \frac{1000 \times 10^6 \ \text{W}}{\left(200 \times 10^6 \ \dfrac{\text{eV}}{\text{fission}}\right)\left(\dfrac{1.602 \times 10^{-19} \text{J}}{1 \ \text{eV}}\right)}$$

$$= \boxed{3.12 \times 10^{19} \ \text{fission/s}}$$

3. Large turbines use impulse control stages because no pressure drop occurs across such a stage, thereby allowing partial arc admission of the steam without significant forces being exerted on the turbine rotor.

The answer is (A).

4.

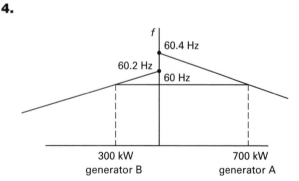

The power carried by an AC generator is

$$P = \frac{f_{n1} - f_{\text{sys}}}{f_{\text{droop}}}$$

Solving for the no-load frequency gives

$$f_{n1} = Pf_{\text{droop}} + f_{\text{sys}}$$

Solving for the no-load frequency increase necessary to carry an additional 300 kW gives

$$\Delta f_{n1} = \Delta P f_{\text{droop}}$$

$$= (300 \ \text{kW})\left(\frac{0.6 \ \text{Hz}}{1000 \ \text{kW}}\right)$$

$$= 0.18 \ \text{Hz}$$

The required no-load frequency is

$$f_{n1} = f_{n1,\text{current}} + \Delta f_{n1}$$

$$= 60.4 \ \text{Hz} + 0.18 \ \text{Hz}$$

$$= \boxed{60.58 \ \text{Hz} \quad (60.6 \ \text{Hz})}$$

Generator B's frequency will have to be lowered by 0.2 Hz to maintain the system frequency at 60 Hz.

5. The real load sharing diagram with generator A carrying all the load is

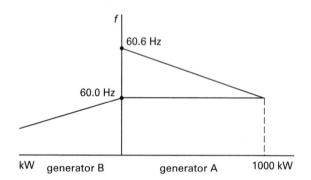

6. The field of a DC machine is normally on the stationary portion, or stator. This allows commutation to be accomplished on the rotor.

The answer is (A).

7. Synchronous speed is derived by

$$n_s = \frac{120 f}{p}$$

The minimum number of poles determines the maximum speed. Two is the minimum number.

$$n_s = \frac{(120)(60 \ \text{Hz})}{2} = \boxed{3600 \ \text{rpm}}$$

8. A transient condition of short duration and high magnitude is called a surge.

The answer is (D).

34 Three-Phase Electricity and Power

PRACTICE PROBLEMS

1. A three-phase 208 V (rms) system supplies heating elements connected in a wye configuration. What is the resistance of each element if the total balanced load is 3 kW?

2. A three-phase system has a balanced delta load of 5000 kW at 84% power factor. If the line voltage is 4160 V (rms), what is the line current?

3. A 120 V (per phase, rms) three-phase system has a balanced load consisting of three 10 Ω resistances. What total power is dissipated if the connection is (a) a wye configuration and (b) a delta configuration?

4. A balanced delta load consists of three 20 Ω∠25° impedances. The 60 Hz line voltage is 208 V (rms). Find the (a) phase current, (b) line current, (c) phase voltage, (d) power consumed by each phase, and (e) total power.

5. A balanced wye load consists of three $3 + j4 \Omega$ impedances. The system voltage is 110 V (rms). Find the (a) phase voltage, (b) line current, (c) phase current, and (d) total power.

6. Consider the one-line diagram of a three-phase distribution system. What is the generator base voltage?

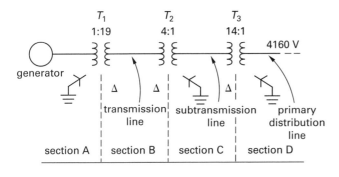

SOLUTIONS

1.

$$R = \frac{V_p^2}{P_p} = \frac{\left(\frac{V_l}{\sqrt{3}}\right)^2}{\frac{P_t}{3}} = \frac{V_l^2}{P}$$

$$= \frac{(208 \text{ V})^2}{3 \times 10^3 \text{ W}}$$

$$= \boxed{14.42 \ \Omega}$$

2.

$$I_l = \left(\frac{1}{\sqrt{3}}\right)\left(\frac{P_\Delta}{V_l \cos \phi}\right)$$

$$= \left(\frac{1}{\sqrt{3}}\right)\left(\frac{5000 \times 10^3 \text{ W}}{(4160 \text{ V})(0.84)}\right)$$

$$= \boxed{826 \text{ A}}$$

3. (a)

$$P_p = \frac{V_p^2}{R} = \frac{(120 \text{ V})^2}{10 \ \Omega} = 1440 \text{ W}$$

$$P_t = 3P_p = (3)(1440 \text{ W}) = \boxed{4320 \text{ W}}$$

(b)

$$P_t = 3\left(\frac{V_l^2}{R}\right) = (3)\left(\frac{(120 \text{ V})^2}{10 \ \Omega}\right)$$

$$= \boxed{4320 \text{ W}}$$

4. (a)

$$\mathbf{I}_p = \frac{\mathbf{V}_l}{\mathbf{Z}} = \frac{208 \text{ V}\angle 0°}{20 \ \Omega\angle 25°}$$

$$= \boxed{10.4 \text{ A}\angle -25°}$$

(b)

$$\mathbf{I}_l = \sqrt{3} \ \angle -30° \ \mathbf{I}_p$$

$$= (\sqrt{3} \ \angle -30°)(10.4 \text{ A}\angle -25°)$$

$$= \boxed{18 \text{ A}\angle -55°}$$

(c)

$$\mathbf{V}_p = \mathbf{V}_l$$

$$= \boxed{208 \text{ V}\angle 0°}$$

(d)

$$P_p = V_p I_p \cos \phi$$

$$= (208 \text{ V})(10.4 \text{ A})\cos 25°$$

$$= \boxed{1960.5 \text{ W}}$$

(e) $\qquad P_\Delta = 3P_p = (3)(1960.5 \text{ W})$

$$= \boxed{5882 \text{ W}}$$

5. (a) $\qquad V_p = \dfrac{110 \text{ V}}{\sqrt{3}} = \boxed{63.51 \text{ V}}$

The impedance is given as $3 + j4\,\Omega$ in rectangular coordinates. The magnitude of the impedance, Z, in complex algebra, is

(b) $\qquad Z = \sqrt{(3\ \Omega)^2 + (4\ \Omega)^2} = 5\ \Omega$

$$\phi = \arctan \dfrac{4\ \Omega}{3\ \Omega} = 53.13°$$

$$\mathbf{I}_l = \dfrac{\mathbf{V}_p}{\mathbf{Z}} = \dfrac{63.51 \text{ V}\angle 0°}{5\ \Omega\angle 53.13°}$$

$$= \boxed{12.70 \text{ A}\angle -53.13°}$$

(c) $\qquad \mathbf{I}_p = \mathbf{I}_l = \boxed{12.70 \text{ A}\angle -53.13°}$

(d) $\qquad P_t = \sqrt{3}\,VI\cos\phi$

$$= \sqrt{3}(110 \text{ V})(12.7 \text{ A})\cos 53.13°$$

$$= \boxed{1451 \text{ W}}$$

6. For a three-phase system, the line voltage is the base. The base in the primary distribution line is given as 4160 V, or $V = 1.0$ pu $\angle 0°$.

Because the transformers scale the base values in each section, the base voltage value of 4160 V will be determined at the generator. The $\sqrt{3}$ is used to account for the wye connection line-to-phase and phase-to-line conversions.

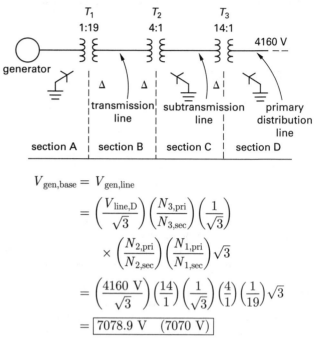

$$V_{\text{gen,base}} = V_{\text{gen,line}}$$

$$= \left(\dfrac{V_{\text{line,D}}}{\sqrt{3}}\right)\left(\dfrac{N_{3,\text{pri}}}{N_{3,\text{sec}}}\right)\left(\dfrac{1}{\sqrt{3}}\right)$$

$$\times \left(\dfrac{N_{2,\text{pri}}}{N_{2,\text{sec}}}\right)\left(\dfrac{N_{1,\text{pri}}}{N_{1,\text{sec}}}\right)\sqrt{3}$$

$$= \left(\dfrac{4160 \text{ V}}{\sqrt{3}}\right)\left(\dfrac{14}{1}\right)\left(\dfrac{1}{\sqrt{3}}\right)\left(\dfrac{4}{1}\right)\left(\dfrac{1}{19}\right)\sqrt{3}$$

$$= \boxed{7078.9 \text{ V} \quad (7070 \text{ V})}$$

This generator line voltage of 7070 V is the 1.0 pu $\angle 0°$ value in section A of the distribution system.

35 Batteries, Fuel Cells, and Power Supplies

PRACTICE PROBLEMS

1. A cell that undergoes a chemical reaction in such a manner that it cannot be recharged is called a

(A) dry cell

(B) primary cell

(C) reserve cell

(D) secondary cell

2. Within a battery, conventional current flows from

(A) anion to anode

(B) anion to cation

(C) anode to cathode

(D) positive electrode to negative electrode

3. A series of fuel cells is connected to produce a no-load voltage of 320 V. At a rated voltage of 300 V the cells provide 2.0 MW to a DC-AC converter. Assuming the cell can be modeled as a standard battery, what is the total internal resistance of the fuel cells?

(A) 0.0001 Ω

(B) 0.003 Ω

(C) 0.05 Ω

(D) 3 Ω

4. A wristwatch mercury cell has a capacity of 200×10^{-3} A·h. How many months can the watch function without replacement if it draws 15×10^{-6} A?

(A) 3 mo

(B) 6 mo

(C) 12 mo

(D) 18 mo

5. Increasing the current withdrawn from a battery has what effect on the battery capacity and why?

(A) increased capacity, which is caused by less internal heat loss

(B) increased capacity, which is caused by lower internal resistance

(C) decreased capacity, which is caused by higher load voltage drop

(D) decreased capacity, which is caused by increased polarization

SOLUTIONS

1. A cell in which the electrodes are consumed and in general cannot be recharged is called a primary cell.

The answer is (B).

2. Within a battery, the conventional current flows from anode to cathode. (The anode is labeled as the positive terminal. Externally, the anode is the terminal into which positive current flows.)

The answer is (C).

3. The circuit is

The voltage across the internal resistance is given (20 V). To find the resistance, the current must be determined. From the rated conditions, the load resistance is

$$P = \frac{V^2}{R_L}$$

$$R_L = \frac{V^2}{P} = \frac{(300 \text{ V})^2}{2.0 \times 10^6 \text{ W}} = 0.045 \ \Omega$$

The rated current is

$$P = I^2 R_L$$

$$I = \sqrt{\frac{P}{R_L}} = \sqrt{\frac{2.0 \times 10^6 \text{ W}}{0.045 \ \Omega}} = 6.67 \times 10^3 \text{ A}$$

Using Ohm's law on the internal resistor gives

$$V = I R_{\text{int}}$$

$$R_{\text{int}} = \frac{V}{I} = \frac{320 \text{ V} - 300 \text{ V}}{6.67 \times 10^3 \text{ A}}$$

$$= \boxed{0.003 \ \Omega}$$

The answer is (B).

4. Amp-hour capacity is

$$Ah = It$$

Therefore,

$$t = \frac{Ah}{I} = \frac{200 \times 10^{-3} \text{ A·h}}{15 \times 10^{-6} \text{ A}}$$

$$= 1.33 \times 10^4 \text{ h}$$

Assuming 30-day months, the capacity is

$$\text{total months} = \frac{1.33 \times 10^4 \text{ h}}{\left(30 \ \frac{\text{d}}{\text{mo}}\right)\left(24 \ \frac{\text{h}}{\text{d}}\right)}$$

$$= \boxed{18.5 \text{ mo}}$$

The answer is (D).

5. Battery capacity drops as the discharge current increases due to increased losses on the internal resistance and polarization of electrodes; that is, hydrogen bubble insulation of the electrodes.

The answer is (D).

36 Power Distribution

PRACTICE PROBLEMS

1. Which of the following voltages is NOT commonly used in the subtransmission portion of the power distribution system?

(A) 25 kV

(B) 69 kV

(C) 115 kV

(D) 138 kV

2. A single-conductor cable with an outside diameter of 0.5 cm is to be used in an underground installation. What is the optimal conductor radius?

(A) 0.05 cm

(B) 0.1 cm

(C) 0.2 cm

(D) 0.3 cm

3. A single-conductor capacitance-graded cable with an outside diameter of 10 cm can withstand a maximum electric field of 700 kV/m before insulation breakdown occurs. What is the maximum operating voltage of the conductor?

4. What is the total capacitance of 100 m of single-conductor cable with the optimal radii ratio and poly-ethylene insulation ($\epsilon_r = 2.25$)?

5. A three-phase 11.5 kV generator drives a 500 kW, 0.866 lagging power factor load. Determine the (a) line current and (b) necessary generator rating.

6. A transmission line with 8% reactance on a 200 MW base connects two substations. At the first substation the voltage is 1.03 pu∠5°, and at the second the voltage is 0.98 pu∠−2.5°. Find the (a) power and (b) volt-amperes reactive flowing out of the first substation.

SOLUTIONS

1. The trend in transmission voltages is toward higher voltages. The most common are 115 kV, 69 kV, 138 kV, and 238 kV. 25 kV is not common.

The answer is (A).

2. The optimal ratio is

$$\frac{r_2}{r_1} = 2.718$$

The outside diameter is given as 0.5 cm. The radius, r_2, is 0.25 cm. Substituting gives

$$r_1 = \frac{r_2}{2.718} = \frac{0.25 \text{ cm}}{2.718} = \boxed{0.0920 \text{ cm} \quad (0.1 \text{ cm})}$$

The answer is (B).

3. The operating voltage for a capacitance-graded cable is

$$V = E_{\max}\left(r_1 \ln \frac{r_2}{r_1} + r_2 \ln \frac{r_3}{r_2}\right)$$
$$= E_{\max}(r_1 \ln \xi + r_2 \ln \xi)$$

The values of r_1 and r_2 are the only unknowns. They are determined from the ratio of radii, ξ.

$$\xi = \frac{r_3}{r_2}$$

$$r_2 = \frac{r_3}{\xi} = \frac{\dfrac{10 \text{ cm}}{2}}{2.718} = 1.84 \text{ cm}$$

$$\xi = \frac{r_2}{r_1}$$

$$r_1 = \frac{r_2}{\xi} = \frac{1.84 \text{ cm}}{2.718} = 0.677 \text{ cm}$$

Substituting known and calculated values gives

$$V = E_{\max}(r_1 \ln \xi + r_2 \ln \xi)$$
$$= \left(700 \ \frac{\text{kV}}{\text{m}}\right)\left(\begin{array}{c}(0.677 \text{ cm})\ln 2.718 \\ + (1.84 \text{ cm})\ln 2.718\end{array}\right)\left(\frac{1 \text{ m}}{100 \text{ cm}}\right)$$

$$= \boxed{17.6 \text{ kV}}$$

4. The capacitance per unit length is

$$C_l = \frac{Q}{V} = \frac{2\pi\epsilon}{\ln\xi} = \frac{2\pi\epsilon_r\epsilon_0}{\ln e}$$

$$= \frac{2\pi(2.25)\left(8.854 \times 10^{-12}\ \frac{F}{m}\right)}{1}$$

$$= 125 \times 10^{-12}\ F/m$$

The total capacitance is

$$C = C_l L$$

$$= \left(125 \times 10^{-12}\ \frac{F}{m}\right)(100\ m)$$

$$= \boxed{125 \times 10^{-10}\ F}$$

5. (a)

$$P_{\text{phase}} = I_p V_p(\text{pf})$$

$$I_p = \frac{P_{\text{phase}}}{V_p(\text{pf})}$$

$$I_l = \left(\frac{P_p}{V_l(\text{pf})}\right)\sqrt{3}$$

$$= \left(\frac{\frac{P_{\text{total}}}{3\ \text{phases}}}{V_l}\right)\sqrt{3}$$

$$= \frac{\frac{500\ kW}{3\ \text{phases}}}{\left(\frac{11.5\ kV}{\sqrt{3}}\right)(0.866)}$$

$$= \boxed{29\ A}$$

(b)

$$P = S(\text{pf})$$

$$S = \frac{P}{\text{pf}}$$

$$= \frac{500\ kW}{0.866}$$

$$= \boxed{577\ kVA}$$

6.

(a)

$$I_{\text{pu}} = \frac{V_{\text{pu}}}{Z_{\text{pu}}} = \frac{1.03\ V\angle 5° - 0.98\ V\angle -2.5°}{j0.08}$$

$$= 1.758\ \text{pu}\angle -19.5°$$

$$S_{\text{pu}} = V_{\text{pu}} I_{\text{pu}}^*$$

$$= (1.03\ \text{pu}\angle 5°)(1.758\ \text{pu}\angle 19.5°)$$

$$= 1.811\ \text{pu}\angle 24.5°$$

$$= 1.648 + j0.751\ \text{pu}$$

$$P = P_{\text{pu}} P_{\text{base}} = (1.648)(200\ MW)$$

$$= \boxed{330\ MW}$$

(b)

$$Q = Q_{\text{pu}} Q_{\text{base}} = (0.751)(200\ MVAR)$$

$$= \boxed{150\ MVAR}$$

37 Power Transformers

PRACTICE PROBLEMS

1. Which electrical element in the exact (real) transformer model accounts for the eddy current and hysteresis losses?

(A) B_c

(B) R_c

(C) R_p

(D) R_s

2. Which electrical element in the exact (real) transformer model accounts for magnetic flux leakage?

(A) B_c

(B) X_p alone

(C) X_s alone

(D) X_p and X_s

3. What is the eddy current power loss for a 500 kVA rated, 200 kg iron-core transformer with a coupling coefficient of 4×10^{-4} in a 1.4 T peak magnetic field?

(A) 3.0 W

(B) 9.0 W

(C) 400 W

(D) 570 W

4. A transformer is rated for 34.5 kV. The measured secondary voltage at no load is 35.02 kV. What is the percent voltage regulation?

(A) 0.50%

(B) 1.4%

(C) 1.5%

(D) 2.0%

5. An open-circuit test is conducted on a 120 V transformer winding. The results are $V_{1oc} = 120$ V, $V_{2oc} = 240$ V, $I_{1oc} = 0.25$ A, and $P_{oc} = 20$ W. What apparent power is necessary to supply core losses in this transformer?

(A) 20 VA

(B) 30 VA

(C) 60 VA

(D) 120 VA

6. A 13.8 kV single-phase transformer is subjected to an open-circuit test with the following results: $P_{in} = 900$ W, $I_{in} = 0.2$ A, and $V_{out} = 460$ V (secondary). Determine the values of the equivalent circuit parameters that can be found from this test.

7. A 32 kVA single-phase transformer is rated at 15 kV on the primary. A 60 Hz short-circuit test on the primary indicates $P_{in} = 1$ kW, $I_{in} = 20$ A, $I_s = 100$ A, and $V_p = 80$ V. Determine the (a) winding resistances and (b) leakage inductances.

8. A single-phase transformer is rated at 115 kV at 500 kVA. A short-circuit test on the high-voltage side at rated current indicates $P_{sc} = 435$ W and $V_{sc} = 2.5$ kV. Determine the per-unit impedance of the transformer.

Distribution

SOLUTIONS

1. Core losses include those caused by eddy currents and hysteresis. In the exact transformer model, these are accounted for by $G_c = 1/R_c$.

The answer is (B).

2. Magnetic flux leakage is a term used to describe any flux that is not mutual. The primary and secondary self-inductances represented by X_p and X_s are not mutual.

The answer is (D).

3. The eddy current loss is found from

$$P_e = k_e B_m^2 f^2 m$$

Assume 60 Hz is the frequency. Substitute the given values.

$$P_e = (4 \times 10^{-4})(1.4 \text{ T})^2 (60 \text{ Hz})^2 (200 \text{ kg})$$
$$= \boxed{565 \text{ W} \quad (570 \text{ W})}$$

The answer is (D).

4. The voltage regulation is

$$\text{VR} = \frac{V_{nl} - V_{fl}}{V_{fl}}$$
$$= \frac{\dfrac{V_p}{a} - V_{s,\text{rated}}}{V_{s,\text{rated}}}$$

Because the voltage was measured at the output of the secondary, it is equivalent to V_p/a. Therefore, the voltage regulation is

$$\text{VR} = \frac{35.02 \text{ kV} - 34.5 \text{ kV}}{34.5 \text{ kV}} \times 100\%$$
$$= \boxed{1.5\%}$$

The answer is (C).

5. The open-circuit test determines the parameters associated with core losses. The apparent power is

$$S_{oc} = I_{1oc} V_{1oc}$$
$$= (0.25 \text{ A})(120 \text{ V})$$
$$= \boxed{30 \text{ VA}}$$

The answer is (B).

6. The open-circuit test obtains the core parameters G and B.

$$G = \frac{P_{oc}}{V_{oc}^2} = \frac{900 \text{ W}}{(13,800 \text{ V})^2} = \boxed{4.73 \times 10^{-6} \text{ S}}$$

$$Q^2 = S^2 - P^2 = \big((13,800 \text{ VA})(0.2 \text{ A})\big)^2 - (900 \text{ W})^2$$

$$Q = 2609 \text{ VAR}$$

$$B = \frac{Q}{V^2} = \frac{2609 \text{ VAR}}{(13,800 \text{ V})^2} = \boxed{13.7 \times 10^{-6} \text{ S}}$$

The turns ratio is

$$\frac{V_p}{N_p} = \frac{V_s}{N_s}$$
$$\frac{N_p}{N_s} = \frac{V_p}{V_s} = \frac{13,800 \text{ V}}{460 \text{ V}}$$
$$= \boxed{30}$$

7. (a)
$$N_p I_p = N_s I_s$$
$$a = \frac{N_p}{N_s} = \frac{I_s}{I_p}$$
$$= \frac{100 \text{ A}}{20 \text{ A}}$$
$$= 5$$
$$P_{sc} = I_p^2 R_p + I_s^2 R_s = 1000 \text{ W}$$
$$I_p^2 (R_p + a^2 R_s) = 1000 \text{ W}$$
$$R_p + a^2 R_s = \frac{1000 \text{ W}}{(20 \text{ A})^2} = 2.5 \text{ }\Omega$$

Assume $R_p = a^2 R_s = \boxed{1.25 \text{ }\Omega.}$

$$R_s = \frac{1.25 \text{ }\Omega}{(5)^2} = \boxed{0.05 \text{ }\Omega}$$

(b)
$$S = V_p I_p = (80 \text{ V})(20 \text{ A}) = 1600 \text{ VA}$$
$$= P + jQ$$
$$Q^2 = S^2 - P^2$$
$$Q = 1249 \text{ VAR}$$
$$= I_1^2 (X_p + a^2 X_s)$$
$$X_p + a^2 X_s = \frac{Q}{I_1^2} = \frac{1249 \text{ VAR}}{(20 \text{ A})^2} = 3.12 \text{ }\Omega$$

Distribution

Assume $X_p = a^2 X_s$, then

$$2X_p = 3.12$$

$$X_p = \frac{3.12}{2} = 1.56 \ \Omega = \omega L_p = 377 L_p$$

$$L_p = \frac{1.56 \ \Omega}{377 \ \frac{\text{rad}}{\text{s}}} = \boxed{4.14 \times 10^{-3} \ \text{H} \quad (4.14 \ \text{mH})}$$

$$X_s = \frac{X_p}{a^2} = \frac{1.56}{a^2} = 0.0624 \ \Omega = \omega L_s = 377 L_s$$

$$L_s = \frac{0.0624 \ \Omega}{377 \ \frac{\text{rad}}{\text{s}}} = \boxed{0.166 \times 10^{-3} \ \text{H} \quad (0.166 \ \text{mH})}$$

8. The per-unit impedance is found from

$$Z_{\text{pu}} = \frac{Z_{\text{actual}}}{Z_{\text{base}}} = \frac{R + jX}{Z_{\text{base}}} \qquad [\text{I}]$$

Find the unknown quantities.

$$Z_{\text{base}} = \frac{V_{\text{base}}}{I_{\text{base}}} \qquad [\text{II}]$$

The short-circuit test is performed at rated current, which is I_{base}.

$$I_{\text{rated}} = I_{\text{base}} = \frac{S_{\text{rated}}}{V_{\text{rated}}}$$

$$= \frac{500 \ \text{kVA}}{115 \ \text{kV}}$$

$$= 4.35 \ \text{A}$$

Substitute the rated current into Eq. II.

$$Z_{\text{base}} = \left(\frac{115 \ \text{kV}}{4.35 \ \text{A}}\right)\left(1000 \ \frac{\text{V}}{\text{kV}}\right) = 26{,}440 \ \Omega$$

Find R from the short circuit (sc) parameters.

$$P_{\text{sc}} = I_{\text{sc}}^2 R = I_{\text{rated}}^2 R$$

$$R = \frac{P_{\text{sc}}}{I_{\text{rated}}^2} = \frac{435 \ \text{W}}{(4.35 \ \text{A})^2} = 23 \ \Omega$$

Find X from

$$Q_{\text{sc}} = I_{\text{rated}}^2 X$$

$$X = \frac{Q}{I_{\text{rated}}^2} \qquad [\text{III}]$$

The reactive power is found using

$$S_{\text{sc}}^2 = P_{\text{sc}}^2 + Q_{\text{sc}}^2$$

$$Q_{\text{sc}} = \sqrt{S_{\text{sc}}^2 - P_{\text{sc}}^2}$$

$$= \sqrt{\left(I_{\text{rated}} V_{\text{sc}}\right)^2 - (435 \ \text{W})^2}$$

$$= \sqrt{\left((4.35 \ \text{A})(2.5 \times 10^3 \ \text{V})\right)^2 - (435 \ \text{W})^2}$$

$$= 10{,}870 \ \text{VAR}$$

Substitute into Eq. III, giving

$$X = \frac{10{,}870 \ \text{VAR}}{(4.35 \ \text{A})^2} = 574 \ \Omega$$

Substitute the calculated values for R, X, and Z_{base} into Eq. I.

$$Z_{\text{pu}} = \frac{23 \ \Omega + j574 \ \Omega}{26{,}440 \ \Omega} = 0.0009 + j0.022 \ \text{pu}$$

$$= \boxed{0.001 + j0.022 \ \text{pu}}$$

Distribution

38 Power Transmission Lines

PRACTICE PROBLEMS

1. As frequency increases, the skin effect causes which of the following parameters to increase?

(A) AC resistance

(B) DC resistance

(C) internal inductance

(D) external inductance

2. What distributed parameter is normally insignificant and ignored in calculations?

(A) line resistance

(B) line inductance

(C) shunt resistance

(D) shunt capacitance

3. The velocity factor is defined as the ratio of the velocity of propagation, v_w, to the speed of light, c. What is the velocity factor in a single waxwing conductor at 60 Hz?

(A) 0.29

(B) 0.61

(C) 0.90

(D) 0.97

4. What is parameter A of a 220 km falcon line with $Z_l = 0.85 \ \Omega/\text{mi} \angle 70°$ and $Y_l = 7 \times 10^{-6} \ \text{S/mi} \angle 90°$?

(A) $4.0 \times 10^{-4} \angle 160°$

(B) $0.948 \angle 1.2°$

(C) $0.996 \angle 0.008°$

(D) $1.0 \angle 20°$

5. A three-phase, 215 kV transmission line supplies a 100 MW wye-connected load at 0.8 pf lagging. The ABCD parameters are $A = 0.9 \angle 2.0°$, $B = 150 \angle 80°$, $C = 0.001 \angle 90°$, and $D = 0.9 \angle 2°$. What is the sending-end voltage?

SOLUTIONS

1. The skin effect causes AC resistance to increase and internal inductance to decrease.

The answer is (A).

2. The shunt resistance is normally very large and has a minimal effect on current and voltage.

The answer is (C).

3. The velocity of wave propagation is

$$v_w = \frac{1}{\sqrt{L_l C_l}}$$

From App. 38.A, a waxwing conductor has an inductive reactance of

$$X_{L,l} = 0.476 \ \Omega/\text{mi}$$

Solving for the inductance per unit length gives

$$X_{L,l} = \omega L = 2\pi f L$$

$$L = \frac{X_{L,l}}{2\pi f} = \frac{0.476 \ \frac{\Omega}{\text{mi}}}{2\pi \ (60 \ \text{Hz})}$$

$$= 1.26 \times 10^{-3} \ \text{H/mi}$$

The capacitive reactance, from App. 38.A, is

$$X_C = 0.1090 \ \text{M}\Omega\text{-mi}$$

Solving for C_l gives

$$|X_C| = \frac{1}{\omega C_l} = \frac{1}{2\pi f C_l}$$

$$C_l = \frac{1}{2\pi f |X_C|} = \frac{1}{2\pi \ (60 \ \text{Hz})(0.1090 \times 10^6 \ \Omega\text{-mi})}$$

$$= 2.43 \times 10^{-8} \ \text{F/mi}$$

Substituting the calculated values gives

$$v_w = \frac{1}{\sqrt{L_l C_l}}$$

$$= \frac{1}{\sqrt{\left(1.26 \times 10^{-3}\ \frac{\text{H}}{\text{mi}}\right)\left(2.43 \times 10^{-8}\ \frac{\text{F}}{\text{mi}}\right)}}$$

$$= \left(1.81 \times 10^5\ \frac{\text{mi}}{\text{s}}\right)\left(1.609\ \frac{\text{km}}{\text{mi}}\right)\left(1000\ \frac{\text{m}}{\text{km}}\right)$$

$$= 2.91 \times 10^8\ \text{m/s}$$

The velocity factor is

$$v_{\text{factor}} = \frac{v_w}{C}$$

$$= \frac{2.91 \times 10^8\ \frac{\text{m}}{\text{s}}}{3.00 \times 10^8\ \frac{\text{m}}{\text{s}}}$$

$$= \boxed{0.97}$$

The answer is (D).

4. This is a medium-length transmission line. Regardless of the model chosen,

$$\text{A} = 1 + \tfrac{1}{2}YZ$$

$$= 1 + \tfrac{1}{2}\left(7 \times 10^{-6}\ \frac{\Omega^{-1}}{\text{mi}}\ \angle 90°\right)\left(0.85\ \frac{\Omega}{\text{mi}}\ \angle 70°\right)$$

$$= 1 + (2.98 \times 10^{-6}\ \text{mi}^{-2} \angle 160°)(220\ \text{km})^2$$

$$\times \left(\frac{1\ \text{mi}}{1.609\ \text{km}}\right)^2$$

$$= 1 \angle 0° + 5.56 \times 10^{-2} \angle 160°$$

$$= (1 + 0j) + (-0.0523 + j0.0190)$$

$$= 0.9477 + j0.0190$$

$$= \boxed{0.948 \angle 1.2°}$$

The type of cable determines the electrical parameters, but these were given. Also, the first term of the parameter, that is, one, can be represented as $1\angle 0°$.

The answer is (B).

5. The sending-end voltage is given by

$$V_S = AV_R + BI_R$$

V_R and I_R must be determined. Because the two-part model represents a single phase, the receiving-end phase voltage for a wye-connected load is

$$V_R = \frac{215\ \text{kV}}{\sqrt{3}} = 124\ \text{kV}$$

Because the power factor is given for the load, let V_R be the reference voltage.

$$V_R = 124\ \text{kV} \angle 0°$$

The receiving-end current is found from

$$P = 3I_R V_R(\text{pf})$$

$$I_R = \frac{P}{3V_R(\text{pf})} = \frac{100 \times 10^6\ \text{W}}{(3)(124 \times 10^3\ \text{V})(0.8)}$$

$$= 336\ \text{A}$$

Because the power factor is 0.8 lagging, the current angle is $-36.8°$.

$$I_R = 336\ \text{A} \angle -36.8°$$

Substituting gives

$$V_S = AV_R + BI_R$$

$$= (0.9 \angle 2.0°)(124 \times 10^3\ \text{V} \angle 0°)$$

$$\quad + (150 \angle 80°)(336\ \text{A} \angle -36.8°)$$

$$= 1.116 \times 10^5\ \text{V} \angle 2.0° + 5.04 \times 10^4\ \text{V} \angle 43.2°$$

$$= \boxed{153\ \text{kV} \angle 14.5°}$$

The power could have been calculated using the quantities, in which case the current would be a function of the power and the line-to-line voltage, V_{ll}.

$$I_R = \frac{P}{\sqrt{3}\,V_{ll}(\text{pf})}$$

39 Power System Analysis

PRACTICE PROBLEMS

1. Consider the following generation system with five nodes and five admittances. In order to use a power software analysis tool to model the distribution system attached to the generator, the equations for each node must be known and then placed in the bus admittance matrix. What is the nodal equation for node 2? How many nodal equations are required to determine all branch currents?

(A) $(V_1 - V_2)\,Y_{21} + (V_2 - V_3)\,Y_{23} + (V_4 - V_2)\,Y_{24} = 0$;
    4 equations

(B) $(V_1 - V_2)\,Y_{21} + (V_2 - V_3)\,Y_{23} + (V_2 - V_4)\,Y_{24} = 0$;
    5 equations

(C) $V_2(-Y_{21} + Y_{23} + Y_{24}) - V_1\,Y_{12} - V_3\,Y_{23} - V_4\,Y_{24} = 0$;
    5 equations

(D) $V_2(Y_{21} + Y_{23} + Y_{24}) - V_1\,Y_{21} - V_3\,Y_{23} - V_4\,Y_{24} = 0$;
    4 equations

2. What is most nearly the value of the apparent power, $\mathbf{S}$, for a single phase circuit for which the voltage is 480 V$\angle 0°$ and the current is 30 A$\angle -37°$?

(A) 0.60 kVA$\angle 37°$

(B) 14 kVA$\angle 37°$

(C) 16 kVA$\angle -37°$

(D) 83 kVA$\angle -37°$

3. The line-to-line voltages of a balanced three-phase system are 120° apart but differ in magnitude and are 30° ahead of the line-to-neutral voltages. If the line-to-neutral voltages are considered to be of unit magnitude, what is the approximate magnitude of the line-to-line voltages, which are also 120° apart?

(A) 1.00 V

(B) 1.41 V

(C) 1.73 V

(D) 2.00 V

4. Which of the following statements regarding symmetrical components is NOT true?

(A) A balanced system has phasor components of equal magnitude.

(B) The method of symmetrical components allows the calculation of unbalanced currents during fault conditions. The calculation uses three separate systems of components that are balanced and equal in magnitude.

(C) In a three-phase system, no zero-sequence line voltages exist except during imbalance or fault conditions.

(D) In delta-connected circuits, zero-sequence currents flow around the delta, and therefore do not affect the line currents.

5. Consider the three-phase transmission system shown. The transmission line current is 62.75 A. The load current from the generator is most nearly what value?

(A) 630 A

(B) 1100 A

(C) 6300 A

(D) 7200 A

SOLUTIONS

1. Write KCL for node 2. Assume all currents flowing out of the node are positive. (Either positive or negative is acceptable, but consistency must be retained for all the nodal equations.)

$$(V_2 - V_1)Y_{21} + (V_2 - V_3)Y_{23} + (V_2 - V_4)Y_{24} = 0$$

Rearrange the equation, expanding the parentheses, and combining like voltage terms.

$$V_2(Y_{21} + Y_{23} + Y_{24}) - V_1 Y_{21} - V_3 Y_{23} - V_4 Y_{24} = 0$$

The first term of this equation is known as the self-admittance of the node. The remaining terms are referred to as the mutual admittances of the node.

The number of nodal equations required to determine all branch currents is one less than the total number of nodes. Because there are five nodes, four node equations are required.

The answer is (D).

2. The apparent power is given by the following equation.

$$\mathbf{S} = VI^*$$

Substitute the given values. Note that the complex conjugate changes the sign of the current angle.

$$\mathbf{S} = VI^* = (480\ \text{V}\angle 0°)(30\ \text{A}\angle -37°)$$
$$= \boxed{14{,}000\ \text{VA}\angle 37° \quad (14\text{k VA}\angle 37°)}$$

The answer is (B).

3. Any two phase-to-neutral phasors (an, bn, or cn) of magnitude 1 with an angle 120° can be combined to obtain the line-to-line quantities (AB, BC, CA). Assume the phasors are located at 0° and 240°. (Consider phase A and phase B in a three-phase system, but with unit magnitude.)

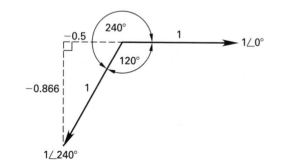

Using KVL around the loop from phase A to neutral, neutral to B, and from the line B to line A, gives the following. (The lower case letters represent phase quantities; the upper case letters represent the desired line

quantity. The notation shown for the result uses the a operator.)

$$V_{an} + V_{nb} + V_{BA} = 0$$
$$-V_{BA} = V_{an} + V_{nb}$$
$$V_{AB} = V_{an} - V_{bn}$$

Now substitute the values for the phase voltage at phases A and B and solve for the line quantity.

$$V_{AB} = 1\ \text{V}\angle 0° - 1\ \text{V}\angle 240°$$
$$= 1\ \text{V}\angle 0° + 1\ \text{V}\angle 60°$$
$$= (1 + j0)\ \text{V} + (0.5 + j0.866)\ \text{V}$$
$$= 1.5\ \text{V} + j0.866\ \text{V}$$
$$= \sqrt{3}\angle 30°\ \text{V}$$
$$= \boxed{1.73\angle 30°\ \text{V} \quad (1.73\ \text{V})}$$

In graphical terms, the result is as shown.

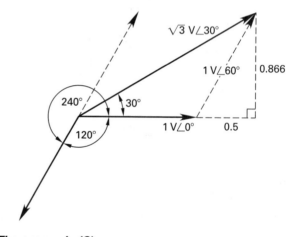

The answer is (C).

4. A balanced system includes components of equal magnitude. Option letter A is true.

The method of symmetrical components was developed to simplify fault calculations. Fortescue's theorem proves three unbalanced phasors can be resolved in a system of three balanced phasors. Therefore, option letter B is true.

Because the voltages of a three phase system always sum to zero, no zero-sequence voltages exist regardless of the imbalance. Option letter C is not true.

In delta-connected circuits, zero-sequence currents flow around the delta and do not impact the line currents. Therefore, option letter D is true.

The answer is (C).

5. The transmission line is downstream of the first transformer whose turns ratio is given by

$$n = \frac{13.2 \text{ kV}}{132 \text{ kV}} = 0.1$$

$$\frac{V_{\text{gen}}}{V_{\text{trans}}} = \frac{I_{\text{trans}}}{I_{\text{gen}}}$$

Solve for the current on the generator side.

$$I_{\text{gen}} = \frac{V_{\text{trans}}}{V_{\text{gen}}} I_{\text{trans}}$$

$$I_{\text{gen}} = \frac{I_{\text{trans}}}{n}$$

$$= \frac{62.75 \text{ A}}{0.1}$$

$$= \boxed{627.5 \text{ A} \quad (630 \text{ A})}$$

The answer is (A).

40 Protection and Safety

PRACTICE PROBLEMS

1. Universal relay characteristics result in the following general diagrams in the R-X plane. Which represent(s) a directional relay?

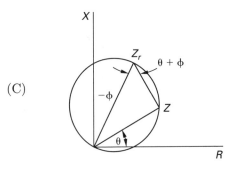

(A)

(B)

(C)

(D) B and C

2. On a schematic of an electrical distribution system, a circle with the number 27 controls a device labeled 52. What does the circle represent?

(A) a time-delay relay operating an undervoltage device

(B) a time-delay relay operating a circuit breaker

(C) an interlocking relay between the two devices

(D) an undervoltage device for an AC circuit breaker

3. An ungrounded system is built for a military marine system with a normal voltage level, A. Following fault analysis design, the insulation of the distribution system is selected. What is most nearly the voltage level for which the insulation on the distribution system is designed?

(A) $(1/\sqrt{3})A$

(B) $\sqrt{2}A$

(C) $\sqrt{3}A$

(D) $2A$

4. A differential relay, designed to protect a zone of the distribution system, uses current transformers (CTs) connected to auxiliary CTs. The auxiliary CTs must match both the phase shift and the turns ratio of the differential relay CTs. If the values do not match, which of the following is the least likely outcome?

(A) The trip could occur at a lower level than expected.

(B) The trip could occur at a higher level than expected.

(C) The trip could occur under non-fault conditions.

(D) The trip will occur only when the CTs are wired in parallel.

5. What is the maximum allowable size of the overcurrent device normally allowed to protect an AWG #12 THHW conductor rated for 90°C used in an ambient temperature of 55°C? Only standard size overcurrent devices are available. Use the NEC.

(A) 10 A

(B) 15 A

(C) 20 A

(D) 30 A

SOLUTIONS

1. An impedance relay is shown in A, a directional relay in B, and a mho relay in C.

The answer is (B).

2. Per the IEEE Standard C37.2 for power system device function, a time-delay relay is indicated by the number 2, an undervoltage relay by 27, an interlocking relay by 3, and an AC circuit breaker by 52. Therefore, a schematic with the numbers 27 and 52 on the device represents an undervoltage device for an AC circuit breaker.

The answer is (D).

3. When an ungrounded system suffers a fault (that is, a ground) on one phase, the other two phases become $\sqrt{3}$ larger than their normal value. (A designer should take the peak value as the "normal" value.) Consider the following phasor diagram showing a grounded phase A.

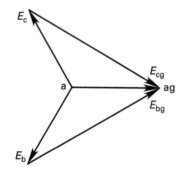

The answer is (C).

4. Depending upon the size of the auxiliary CTs, a trip may occur lower or higher than expected. Further, the trip could occur under normal (that is, non-fault) conditions. Answers (A), (B), and (C) are possible outcomes.

The differential and auxiliary CTs will not serve their function if wired in parallel.

The answer is (D).

5. The overcurrent protection for conductors is specified by NEC Article 240.4, which requires the overcurrent protection to be in accordance with the ampacity of the conductors as specified in NEC Article 310.15, unless otherwise permitted by NEC Article 240.4(A) through Article 240.4(G).

NEC Article 310.15 refers to NEC Table 310.16 to determine ampacity. According to NEC Table 310.16, a THHW sized at AWG #12 and rated for 90°C is rated for 30 A. The ambient temperature assumption for the table is 30°C. The correction factor for an ambient temperature of 55°C is 0.76. The maximum amperage allowed is

$$\text{ampacity} = (\text{rating})(\text{correction factor})$$
$$= (30 \text{ A})(0.76)$$
$$= 22.8 \text{ A}$$

However, per NEC Article 240.4(D)(5), the maximum overcurrent protection cannot exceed 20 A for an AWG #12 wire, unless the system qualifies for the exceptions. The system described does not qualify for the exceptions, so even though the conductor is rated to endure 22.8 A, NEC requires that overcurrent protection not exceed 20 A. If the calculation of the ampacity had been a number smaller than 20 A, NEC would require use of overcurrent protection not exceeding that number.

The answer is (C).

Protection and Safety

41 Rotating DC Machinery

PRACTICE PROBLEMS

1. A four-pole generator is turned at 3600 rpm. Each of its poles has a flux of 3.0×10^{-4} Wb. The armature has a total of 200 conductors and is simplex lap wound. What average voltage is generated at no load?

2. A DC generator has an armature resistance of 0.31 Ω. The shunt field resistance is 134 Ω. No-load voltage is 121 V at 1775 rpm. Full-load current is 30 A at 110 V. Assume a constant flux. What are the (a) no-load shunt field current, (b) full-load speed, and (c) copper power losses at full load?

3. (a) What is the back emf of a 10 hp shunt motor operating on 110 V terminals if 90 A flow through a 0.05 Ω armature resistance? (b) What is the line current if the field resistance is 60 Ω?

4. The nameplate of a 240 V DC motor states that the full-load line current is 67 A. During a no-load test at rated voltage, 3.35 A armature current and 3.16 A field current are drawn. The no-load armature resistance is 0.207 Ω. The no-load brush drop is 2 V. What are the (a) horsepower and (b) efficiency at full load?

5. A DC generator has a 250 V output at a 50 A load current and a 260 V output at no load. Estimate the armature resistance.

6. A separately excited DC generator is rated at 10 kW, 240 V, and 2800 rpm for an input power of 15 hp. The no-load voltage is 260 V. Determine the (a) armature resistance and (b) mechanical power loss at 2800 rpm.

7. A DC shunt motor is running at rated values when the line voltage suddenly drops by 10%. Determine the resulting changes in input power and motor speed if (a) the load torque is constant and (b) the load torque is proportional to the speed.

8. A DC series motor is running at rated speed, voltage, and current when the load torque increases by 25%. Determine the effects of this change, assuming that the line voltage remains the same.

9. A DC shunt motor receives 200 V and 25 A at 1000 rpm and 5 hp. It has an armature resistance of 0.5 Ω and a shunt field resistance of 100 Ω. Determine the (a) rotational losses and (b) no-load speed at 200 V.

10. A DC shunt motor is rated at 7.5 hp, 60 A, 1000 rpm, and 120 V. Assume the load torque is constant, regardless of speed. What field resistance values are required to change the speed to (a) 1150 rpm and (b) 750 rpm?

11. Consider the following figure. Which line represents the characteristics of a series motor?

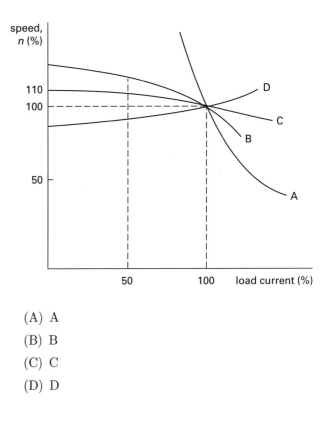

(A) A

(B) B

(C) C

(D) D

SOLUTIONS

1. For z conductors, a parallel armature paths, p poles, a magnetic flux, Φ, and a rotational speed, n,

$$V = \left(\frac{z}{a}\right)p\Phi\left(\frac{n}{60}\right)$$

For simplex lap winding, $a = p$.

$$V = \left(\frac{200}{4}\right)(4)(3.0 \times 10^{-4}\text{ Wb})\left(\frac{3600\ \frac{\text{rev}}{\text{min}}}{60\ \frac{\text{sec}}{\text{min}}}\right)$$

$$= \boxed{3.6\text{ V}}$$

2. (a) The parameters pertain to full load, fl; no load, nl; field, f; and armature, a.

$$I_{f,\text{nl}} = \frac{V_{\text{nl}}}{R_f} = \frac{121\text{ V}}{134\ \Omega}$$

$$= \boxed{0.903\text{ A}}$$

(b) Voltage is proportional to rotational speed.

$$n_{\text{fl}} = \left(\frac{V_{\text{fl}} + R_a\left(I_{\text{fl}} + \frac{V_{\text{fl}}}{R_f}\right)}{V_{\text{nl}} + R_a\left(\frac{V_{\text{nl}}}{R_f}\right)}\right)n_{\text{nl}}$$

$$= \left(\frac{110\text{ V} + (0.31\ \Omega)\left(30\text{ A} + \frac{110\text{ V}}{134\ \Omega}\right)}{121\text{ V} + (0.31\ \Omega)\left(\frac{121\text{ V}}{134\ \Omega}\right)}\right)\left(1775\ \frac{\text{rev}}{\text{min}}\right)$$

$$= \boxed{1750\text{ rev/min (1750 rpm)}}$$

(c) $$P = \frac{V_{\text{fl}}^2}{R_f} + \left(I_{\text{fl}} + \frac{V_{\text{fl}}}{R_f}\right)^2 R_a$$

$$= \frac{(110\text{ V})^2}{134\ \Omega} + \left(30\text{ A} + \frac{110\text{ V}}{134\ \Omega}\right)^2(0.31\ \Omega)$$

$$= \boxed{385\text{ W}}$$

3. (a) $$E = V_{\text{fl}} - I_{a,\text{load}}R_a$$
$$= 110\text{ V} - (90\text{ A})(0.05\ \Omega)$$
$$= \boxed{105.5\text{ V}}$$

(b) $$I_{\text{fl}} = \frac{V_{\text{fl}}}{R_f} + I_{a,\text{load}} = \frac{110\text{ V}}{60\ \Omega} + 90\text{ A}$$
$$= \boxed{91.8\text{ A}}$$

4.

The total no-load power input to the armature is

$$P_a = I_a V = (3.35\text{ A})(240\text{ V})$$
$$= 804\text{ W}$$

The I^2R power loss in the armature resistance is

$$P_{\text{resistance}} = I_a^2 R = (3.35\text{ A})^2(0.207\ \Omega)$$
$$= 2.323\text{ W}$$

The armature brush power loss is

$$P_{\text{brush}} = I_a V_{\text{brush}}$$
$$= (3.35\text{ A})(2\text{ V})$$
$$= 6.7\text{ W}$$

The no-load rotational loss is

$$P_{\text{rotational}} = P_a - P_{\text{resistance}} - P_{\text{brush}}$$
$$= 804\text{ W} - 2.323\text{ W} - 6.7\text{ W}$$
$$= 794.9\text{ W} \quad (795.0\text{ W})$$

At full load, the total input power is

$$P_{\text{in}} = I_{\text{fl}} V_{\text{nameplate}}$$
$$= (67\text{ A})(240\text{ V})$$
$$= 16,080\text{ W}$$

The field current is the same as for the no-load condition.

$$I_f = 3.16\text{ A}$$

The field power loss is

$$P_f = I_f V = (3.16 \text{ A})(240 \text{ V})$$
$$= 758.4 \text{ W}$$
$$I_a = I_{fl} - I_f = 67 \text{ A} - 3.16 \text{ A}$$
$$= 63.84 \text{ A}$$
$$P_{\text{resistance}} = I_a^2 R = (63.84 \text{ A})^2 (0.207 \text{ }\Omega)$$
$$= 843.6 \text{ W}$$

Using the no-load data, the equivalent brush resistance is

$$R_{\text{brush}} = \frac{V_{\text{brush}}}{I} = \frac{2 \text{ V}}{3.35 \text{ A}} = 0.597 \text{ }\Omega$$

The armature brush power loss is

$$P_{\text{brush}} = I_a^2 R_{\text{brush}} = (63.84 \text{ A})^2 (0.597 \text{ }\Omega)$$
$$= 2433.1 \text{ W}$$

The rotational loss is the same as for the no-load condition.

$$P_{\text{rotational}} = 795.0 \text{ W}$$

The net power to the load is

$$P_{\text{net}} = P_{\text{in}} - P_f - P_{\text{resistance}} - P_{\text{brush}} - P_{\text{rotational}}$$
$$= 16{,}080 \text{ W} - 758.4 \text{ W} - 843.6 \text{ W}$$
$$\quad - 2433.1 \text{ W} - 795.0 \text{ W}$$
$$= 11{,}249.9 \text{ W}$$

(a) The motor output horsepower is

$$P = \frac{(11{,}249.9 \text{ W})\left(1.341 \dfrac{\text{hp}}{\text{kW}}\right)}{1000 \dfrac{\text{W}}{\text{kW}}}$$
$$= \boxed{15.09 \text{ hp}}$$

(b) The efficiency is

$$\eta = \frac{P_{\text{net}}}{P_{\text{in}}}$$
$$= \frac{11{,}249.9 \text{ W}}{16{,}080 \text{ W}}$$
$$= \boxed{0.70 \quad (70.0\%)}$$

5.

$$V_t = E_g - R_a I_{\text{load}}$$

At no load,

$$V_t = E_g = 260 \text{ V}$$

At full load,

$$V_t = E_g - (50 \text{ A}) R_a$$
$$250 \text{ V} = E_g - (50 \text{ A}) R_a$$
$$R_a = \frac{V_{\text{nl}} - V_{\text{fl}}}{I_{\text{fl}}}$$
$$= \frac{260 \text{ V} - 250 \text{ V}}{50 \text{ A}}$$
$$= \boxed{0.2 \text{ }\Omega}$$

6. (a)
$$I_{\text{fl}} = \frac{10 \times 10^3 \text{ W}}{240 \text{ V}} = 41.67 \text{ A}$$
$$R_a = \frac{V_{\text{nl}} - V_{\text{fl}}}{I_{\text{fl}}} \quad [\text{see Prob. 41.5}]$$
$$= \frac{260 \text{ V} - 240 \text{ V}}{41.67 \text{ A}}$$
$$= \boxed{0.48 \text{ }\Omega}$$

(b)
$$P_{\text{in}} = P_{\text{mech. loss}} + I_a^2 R_a + p_{\text{out}}$$
$$= (15 \text{ hp})\left(745.7 \dfrac{\text{W}}{\text{hp}}\right)$$
$$= 11.186 \times 10^3 \text{ W} \quad (11.186 \text{ kW})$$
$$I_a^2 R_a = (41.67 \text{ A})^2 (0.48 \text{ }\Omega)$$
$$= 0.833 \times 10^3 \text{ W} \quad (0.833 \text{ kW})$$
$$P_{\text{out}} = 10 \times 10^3 \text{ W} \quad (10 \text{ kW})$$
$$P_{\text{mech. loss}} = 11.186 \times 10^3 \text{ W} - 10.833 \times 10^3 \text{ W}$$
$$= \boxed{353 \text{ W}}$$

7. For field flux, Φ is proportional to line voltage.

$$\Phi = aV$$

$$\Omega_r = \text{rotor speed} \ [\text{rad/s}]$$

$$E_g = k\Phi\Omega_r$$

$$= kaV\Omega_r$$

$$T = k\Phi I_a = kaVI_a \quad [\text{torque in N·m}]$$

$$V = E_g + R_a I_a$$

$$P_{\text{in}} \approx VI_a$$

$$P_{\text{out}} \approx T\Omega_r$$

The parameters pertain to the field, f; armature, a; generator, g; and rotor, r.

(a) When V is decreased by 10%, to keep T constant, I_a is increased by approximately 10%, and E_g is decreased by 10% with V.

$$P_{\text{in}} = \boxed{VI_a \approx \text{ constant}}$$

$$P_{\text{out}} = P_{\text{in}} - I_a^2 R_a - \text{mechanical loss}$$

Because I_a has increased by 10%, I_a^2 is increased by 20%, but $I_a^2 R_a$ is on the order of 5% of the load power, so it only causes a 1% drop in P_{out}.

E_g drops 10% with V, but

$$\boxed{\Omega_r \text{ drops only 1\%.}}$$

(b) For $T \propto \Omega_r$, in a first approximation,

$$E_g \approx V$$

$$E_g \approx kaV\Omega_r$$

$ka\Omega_r \approx 1$, and there is no change in speed.

In a second approximation,

$$I_a = \frac{V - E_g}{R_a}$$

$$T = kaVI_a$$

The air-gap power is

$$P_{\text{air gap}} = E_g I_a = T\Omega_r$$

$$T = m\Omega_r$$

$$T\Omega_r = m\Omega_r^2$$

$$E_g = kaV\Omega_r$$

$$E_g I_a = \frac{kaV^2\Omega_r(1 - ka\Omega_r)}{R_a} = m\Omega_r^2$$

$$\frac{kaV^2}{R_a} = \frac{m\Omega_r}{1 - ka\Omega_r} = \frac{m}{\frac{1}{\Omega_r} - ka}$$

When V goes down 10%, $1/\Omega_r - ka$ must go up by 21%, so Ω_r must decrease some amount. If under normal conditions $I_a R_a = 0.05V$,

$$I_a R_a = V - E_{g,\text{nl}} = V(1 - ka\Omega_{r,\text{nl}})$$

$$1 - ka\Omega_{r,\text{nl}} = 0.05$$

$$ka = \frac{0.95}{\Omega_{r,\text{nl}}}$$

When V goes down 10%,

$$\frac{m\Omega_r}{1 - ka\Omega_r} \rightarrow (0.81)\left(\frac{m\Omega_{r,\text{nl}}}{1 - ka\Omega_{r,\text{nl}}}\right)$$

$$\frac{\Omega_r}{1 - (0.95)\left(\frac{\Omega_r}{\Omega_{r,\text{nl}}}\right)} = \left(\frac{0.81}{0.05}\right)\Omega_{r,\text{nl}} = 16.2\Omega_{r,\text{nl}}$$

$$\frac{\Omega_r}{\Omega_{r,\text{nl}}} = (16.2)\left(1 - (0.95)\left(\frac{\Omega_r}{\Omega_{r,\text{nl}}}\right)\right)$$

$$= \frac{16.2}{1 + (0.95)(16.2)}$$

$$= 0.99$$

$$\boxed{\Omega_r \text{ drops by about 1\%.}}$$

$$I_a R_a = V - E_g = V - ka\Omega_r$$

$$T = m\Omega_r = kaVI_a$$

Eliminating Ω_r gives

$$I_a R_a = \frac{V}{1 + \frac{(kaV)^2}{mRa}}$$

The original value of $I_a R_a$ is

$$I_{a,\text{nl}} R_a \approx 0.05 V_{\text{nl}}$$

V_0 is the initial voltage. Then,

$$1 + \frac{(ka\,V_{\text{nl}})^2}{mR_a} \approx 20$$

When the voltage drops 10%,

$$1 + \frac{(ka(0.9)\,V_{\text{nl}})^2}{mR_a} = 16.39$$

The resulting armature IR drop is

$$I_aR_a = \frac{0.9\,V_{\text{nl}}}{16.39} = 0.0549\,V_{\text{nl}}$$

Dropping the common V_{nl} term, the current changes by

$$\frac{I_aR_a - I_{a,\text{nl}}R_a}{I_{a,\text{nl}}R_a} = \frac{0.0549 - 0.05}{0.05}$$

$$= 0.098 \text{ pu}$$

The input power changes also.

$$P_{\text{in}} = (0.9\,V_{\text{nl}})(1.098\,I_{a,\text{nl}}) = 0.988P_{\text{nl}}$$

$$\boxed{\text{The power drops by about 1\%.}}$$

8.

$$V = RI + E_g$$
$$E_g = k\Phi\Omega_r = kLI\Omega_r$$
$$T = k\Phi I = kLI^2$$

If T is up 25%, so is I^2, and $\sqrt{1.25} = 1.11$, so

$$\boxed{I \text{ is up 11\%.}}$$

Using subscript zero for the initial values,

$$E_{g0} = V - I_0R_a$$

When I increases by 11%, using subscript 1 for this condition,

$$\boxed{E_g \text{ decreases by about 1\%.}}$$

$$E_{g1} = 0.99E_{g0}$$
$$E_{g0} = kLI_0\Omega_{r0}$$
$$E_{g1} = 0.99kLI_0\Omega_{r0}$$
$$= kLI_1\Omega_{r1}$$
$$\Omega_{r1} = \left(\frac{0.99I_0}{I_1}\right)\Omega_{r0}$$
$$= \left(\frac{0.99}{1.11}\right)\Omega_{r0}$$
$$= 0.89\Omega_{r0}$$

$$\boxed{\text{The speed also drops by about 11\%.}}$$

The output power is

$$P_2 = (1.25\,T_0)(0.89\Omega_{r0}) = 1.11P_0$$

$$\boxed{\text{The air-gap power increases by 11\%.}}$$

9. (a)
$$I_t = 25 \text{ A} = I_a + I_f$$
$$I_f = \frac{V}{R} = \frac{200 \text{ V}}{100 \ \Omega} = 2 \text{ A}$$
$$I_a = I_t - I_f = 25 \text{ A} - 2 \text{ A} = 23 \text{ A}$$
$$E_g = V - I_aR_a = 200 \text{ V} - (23 \text{ A})R_a = 188.5 \text{ V}$$
$$P_{\text{air gap}} = E_gI_a = (188.5 \text{ V})(23 \text{ A})$$
$$= 4335.5 \text{ W}$$
$$P_{\text{load}} = (5 \text{ hp})\left(745.7 \ \frac{\text{W}}{\text{hp}}\right) = 3728.5 \text{ W}$$
$$P_{\text{rotational losses}} = P_{\text{air gap}} - P_{\text{load}} = 4335.5 \text{ W} - 3728.5 \text{ W}$$
$$= \boxed{607 \text{ W}}$$

(b) The speed, n, is proportional to the counter voltage as seen from

$$E_g = k\Phi n$$

The term $k\Phi$ is determined from the information in Part (a) to be

$$k\Phi = \frac{188.5 \text{ V}}{1000 \ \frac{\text{rev}}{\text{min}}}$$

Assuming the rotational losses are fixed at 607 W,

$$E_g I_a = 607 \text{ W}$$
$$I_a = \frac{607 \text{ W}}{E_g}$$

The armature current at the specified 200 V is

$$I_a = \frac{200 \text{ V} - E_g}{0.5 \text{ }\Omega}$$

Substitute for the armature current, giving

$$\frac{607 \text{ W}}{E_g} = \frac{200 \text{ V} - E_g}{0.5 \text{ }\Omega}$$

Rearranging gives the quadratic equation

$$E_g^2 - 200 E_g + 303.5 = 0$$

Solving the quadratic and using the positive sign gives

$$E_g = \frac{-(-200 \text{ V}) \pm \sqrt{(-200 \text{ V})^2 - (4)(1)(303.5 \text{ V}^2)}}{(2)(1)}$$
$$= 198.5 \text{ V}$$

Comparing voltage to rpm ratios gives

$$\frac{n}{198.5 \text{ V}} = \frac{1000 \frac{\text{rev}}{\text{min}}}{188.5 \text{ V}}$$
$$n = \boxed{1053 \text{ rev/min} \quad (1053 \text{ rpm})}$$

10.

$$P_{\text{in}} = V_a I_{\text{in}} = (60 \text{ A})(120 \text{ V}) = 7200 \text{ W}$$
$$P_{\text{out}} = (7.5 \text{ hp})\left(745.7 \frac{\text{W}}{\text{hp}}\right) = 5593 \text{ W}$$

Because load torque is constant,

$$P_{\text{out}} = k_T n$$
$$k_T = \frac{5593 \text{ W}}{1000 \frac{\text{rev}}{\text{min}}}$$
$$= 5.593n \quad [n \text{ in rev/min}]$$

The losses are the difference between P_{in} and P_{out}.

$$P_{\text{loss}} = P_{\text{in}} - P_{\text{out}} = 7200 \text{ W} - 5593 \text{ W}$$
$$= 1607 \text{ W}$$

Because no information is given, assign these losses to the rotational losses, the armature losses $I_a^2 R_a$, and the field losses $I_f^2 R_f$, making all three equal in magnitude.

$$I_a^2 R_a = I_f^2 R_f = P_{\text{rotational losses}} = 535 \text{ W}$$
$$I_f^2 R_f = \frac{V_a^2}{R_f} = 535 \text{ W}$$
$$R_f = \frac{(120 \text{ V})^2}{535 \text{ W}} = 26.9 \text{ }\Omega$$
$$I_f = \frac{120 \text{ V}}{26.9 \text{ }\Omega} = 4.46 \text{ A}$$

Therefore,

$$I_a = I_{\text{in}} - I_f = 60 \text{ A} - 4.46 \text{ A} = 55.54 \text{ A}$$
$$R_a = \frac{P_{\text{rotational losses}}}{I_a^2} = \frac{535 \text{ W}}{(55.54 \text{ A})^2} = 0.173 \text{ }\Omega$$

Because of rounding, the powers are recalculated and the remainder assigned to rotational losses.

$$P_{\text{rotational losses}} = P_{\text{loss}} - \frac{V_a^2}{R_f} - R_a I_a^2$$
$$= 1607 \text{ W} - \frac{(120 \text{ V})^2}{26.9 \text{ }\Omega}$$
$$- (0.173 \text{ }\Omega)(55.54 \text{ A})^2$$
$$= 538 \text{ W}$$

The air-gap power must then be

$$P_{\text{air gap}} = P_{\text{out}} + P_{\text{rotational losses}} = 5.593n + 538 \text{ W}$$

The air-gap power is also $E_g I_a$, and E_g is proportional to I_f and n.

$$E_g = k' I_f n = k'' \left(\frac{n}{R_f}\right)$$

At $n = 1000$ rpm and $R_f = 26.9 \ \Omega$,

$$E_g = 120 \text{ V} - (55.54 \text{ A})(0.173 \ \Omega)$$

$$= 110.4 \text{ V}$$

$$k'' = \frac{(110.4 \text{ V})(26.9 \ \Omega)}{1000 \ \frac{\text{rev}}{\text{min}}}$$

$$= 2.97 \ \frac{\text{V} \cdot \Omega}{\frac{\text{rev}}{\text{min}}}$$

$$E_g = 2.97 \left(\frac{n}{R_f} \right)$$

$$I_a = \frac{120 \text{ V} - E_g}{R_a}$$

$$= \frac{120 \text{ V} - (2.97) \left(\dfrac{n}{R_f} \right)}{0.173 \ \Omega}$$

$$= 694 \text{ A} - (17.2) \left(\frac{n}{R_f} \right)$$

$$P_{\text{air gap}} = E_g I_a$$

$$= (2061) \left(\frac{n}{R_f} \right) - (51.08) \left(\frac{n}{R_f} \right)^2$$

Equating with the mechanical side,

$$(51.08) \left(\frac{n}{R_f} \right)^2 + \left(5.593 - \frac{2061}{R_f} \right) n + 538 \text{ W} = 0$$

$$0 = R_f^2 - \left(\frac{2061 n}{5.593 n + 538 \text{ W}} \right) (R_f) + \frac{51.08 n^2}{5.593 n + 538 \text{ W}}$$

(a) For $n = 1150$ rpm,

$$R_f^2 - (340.0 \ \Omega) R_f + 9692 \ \Omega^2 = 0 \ \Omega^2$$

$$R_f = \boxed{31.4 \ \Omega}$$

(b) For $n = 750$ rpm,

$$R_f^2 - (326.6 \ \Omega) R_f + 6071 \ \Omega^2 = 0 \ \Omega^2$$

$$R_f = \boxed{19.8 \ \Omega}$$

11. A series motor generates large torque at low speed (high current means high torque) and overspeeds at low loads (low current means low torque, but with little or no counter electromagnetic force to counteract the applied terminal voltage, the motor overspeeds). The graph line that shows those properties is A.

The answer is (A).

42 Rotating AC Machinery

PRACTICE PROBLEMS

1. An old single-phase generator has four poles. The armature is simplex lap wound and has a total of 240 armature conductors. The effective armature length and radius are 16.7 cm and 5 cm, respectively. The armature rotates at 1800 rpm. (a) If the uniform magnetic flux density per pole is 1.0 T, what is the maximum voltage induced? (b) What is the horsepower of the driving motor if the rated output is 1 kW and the conversion efficiency is 90%?

2. (a) What is the rotational speed of a 24-pole alternator that produces a 60 Hz sinusoid? (b) If the effective emf is 2200 V and each phase coil has 20 turns in series, what is the maximum flux?

3. Calculate the full-load phase current drawn by a 440 V (rms), 20 hp (per phase) induction motor having a full-load efficiency of 86% and a full-load power factor of 76%.

4. A 200 hp, three-phase, four-pole, 60 Hz, 440 V (rms) squirrel-cage induction motor operates at full load with an efficiency of 85%, a power factor of 91%, and 3% slip. Find the (a) speed in rpm, (b) torque developed, and (c) line current.

5. A factory's induction motor load draws 550 kW at 82% power factor. What size synchronous motor is required to carry 250 hp and raise the power factor to 95%? The line voltage is 220 V (rms).

6. The nameplate of an induction motor lists 960 rpm as the full-load speed. For what frequency was the motor designed?

7. A three-phase, 440 V (line-to-line) synchronous motor has a synchronous reactance (X_s) of 20 Ω and an armature resistance of 0.5 Ω per phase. The synchronous speed is 1800 rpm. The no-load field current is adjusted for minimum armature current at a field current (I_f) of 5 A DC, at which time the input power is 900 W. This machine is to provide 10 kVAR of power factor correction when running lightly loaded. Determine the rotor current required.

8. A 440 V, three-phase synchronous motor is driving a 45 hp pump. Assume the motor is 93% efficient. The rotor current is adjusted until the line current is at a minimum, occurring when the rotor current is 8.2 A DC. Determine the rotor current necessary to provide a leading power factor of 0.8.

9. A 50 hp, 50 kVA, 1746 rpm squirrel-cage induction motor has phase windings rated at 440 V. With the windings connected in a delta configuration, a blocked rotor test is carried out under rated current conditions. The result in per-phase volt-amperes is found to be $P + jQ = 1097 + j\,2791$ VA. This machine is fed from a three-phase, 50 kVA, 440 V transformer that has 10% reactance per phase (on its own base). Determine the dip in the line voltage when the motor is started. The transformer can be modeled as a wye connection of 254 V sources in series with the 10% reactance on each phase.

10. A 50 hp, 50 kVA, 1746 rpm squirrel-cage induction motor has phase windings rated at 440 V and is normally connected in delta. Its starting current is 367 A and 38.3 kW per phase when started directly across the line, and under those conditions it develops a starting torque of 190 ft-lbf. A blocked rotor test at the rated current results in a voltage of 78.6 V and a power of 1.1 kW per phase. A no-load test at the rated voltage results in a current of 25.9 A and 1.1 kW per phase. If this machine is started in the wye configuration, determine its starting current.

11. What is the synchronous speed of a two-pole 400 Hz generator?

    (A) 120 rpm

    (B) 200 rpm

    (C) 7200 rpm

    (D) 24,000 rpm

SOLUTIONS

1. (a) The maximum voltage is

$$V_m = \frac{\pi}{2} V_{\text{ave}}$$

In terms of the rotational speed, n, the maximum voltage is given in Table 42.1, using the total number of poles, p; the number, N, of series armature paths; area, A; and magnetic flux density, B.

$$V_m = \frac{\pi p n N A B}{60}$$

$$= \pi p \left(\frac{n}{60}\right)\left(\frac{z}{a}\right) A B$$

For a simplex lap-wound armature, $a = p$.

$$V_m = 4\pi \left(\frac{1800 \; \frac{\text{rev}}{\text{min}}}{60 \; \frac{\text{s}}{\text{min}}}\right)\left(\frac{240}{4}\right)$$

$$\times (0.10 \text{ m})(0.167 \text{ m})\left(1.0 \; \frac{\text{Wb}}{\text{m}^2}\right)$$

$$= \boxed{377.8 \text{ V}}$$

(b) $$P_{\text{in}} = \frac{P_{\text{out}}}{\text{efficiency}} = \frac{(1 \text{ kW})\left(\frac{1 \text{ hp}}{0.7457 \text{ kW}}\right)}{0.9}$$

$$= \boxed{1.49 \text{ hp}}$$

2. (a) $$n = 120 \frac{f}{p}$$

$$n = \left(2 \; \frac{\text{pole·rev}}{\text{cycle}}\right)\left(60 \; \frac{\text{s}}{\text{min}}\right)\left(\frac{f}{p}\right)$$

$$= \left(2 \; \frac{\text{pole·rev}}{\text{cycle}}\right)\left(60 \; \frac{\text{s}}{\text{min}}\right)\left(\frac{60 \text{ Hz}}{24 \text{ poles}}\right)$$

$$= \boxed{300 \text{ rpm}}$$

(b) $$BA = \frac{\sqrt{2} V_{\text{eff}}}{N\omega} = \frac{\sqrt{2}(2200 \text{ V})}{(20)(2\pi)(60 \text{ Hz})}$$

$$= \boxed{0.413 \text{ Wb}}$$

3.

$$I_p = \frac{P_p}{\eta V_p \cos \phi}$$

$$= \frac{(20 \text{ hp})\left(745.7 \; \frac{\text{W}}{\text{hp}}\right)}{(0.86)(440 \text{ V})(0.76)}$$

$$= \boxed{51.86 \text{ A}}$$

4. (a) $$n_r = \left(\frac{2f}{p}\right)(1-s)$$

$$= \left(\frac{(2)\left(60 \; \frac{\text{s}}{\text{min}}\right)(60 \text{ Hz})}{4}\right)(1 - 0.03)$$

$$= \boxed{1746 \text{ rpm}}$$

(b) $$T = \frac{P}{\Omega} = \frac{(200 \text{ hp})\left(550 \; \frac{\text{ft-lbf}}{\text{s}}{\text{hp}}\right)}{2\pi\left(1746 \; \frac{\text{rev}}{\text{min}}\right)\left(\frac{1 \text{ min}}{60 \text{ s}}\right)}$$

$$= \boxed{602 \text{ ft-lbf}}$$

(c) $$I_l = \left(\frac{1}{\sqrt{3}}\right)\left(\frac{P}{\eta V_l \cos \phi}\right)$$

$$= \left(\frac{1}{\sqrt{3}}\right)\left(\frac{(200 \text{ hp})\left(745.7 \; \frac{\text{W}}{\text{hp}}\right)}{(0.85)(440 \text{ V})(0.91)}\right)$$

$$= \boxed{253 \text{ A}}$$

5.

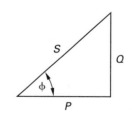

The original power angle is

$$\phi_1 = \arccos 0.82 = 34.92°$$
$$P_1 = 550 \text{ kW}$$
$$Q_i = P_1 \tan \phi_1$$
$$= (550 \text{ kW})\tan 34.92°$$
$$= 384.0 \text{ kVAR}$$

The new conditions are

$$P_2 = (250 \text{ hp})\left(0.7457 \frac{\text{kW}}{\text{hp}}\right)$$
$$= 186.4 \text{ kW}$$
$$\phi_f = \arccos 0.95 = 18.19°$$

Because both motors perform real work,

$$P_f = P_1 + P_2 = 550 \text{ kW} + 186.4 \text{ kW}$$
$$= 736.4 \text{ kW}$$

The new reactive power is

$$Q_f = P_f \tan \phi_f = (736.4 \text{ kW})\tan 18.19°$$
$$= 242.0 \text{ kVAR}$$

The change in reactive power is

$$\Delta Q = Q_i - Q_f$$
$$= 384.0 \text{ kVAR} - 242.0 \text{ kVAR} = 142 \text{ kVAR}$$

Synchronous motors used for power factor correction are rated by apparent power.

$$S = \sqrt{(\Delta P)^2 + (\Delta Q)^2}$$
$$= \sqrt{(186.4 \text{ kW})^2 + (142 \text{ kVAR})^2}$$
$$= \boxed{234.3 \text{ kVA}}$$

6.
$$f = \frac{pn_s}{120}$$
$$= \frac{pn}{120(1-s)}$$

The slip and number of poles are unknown. Assume $s = 0$ and $p = 4$.

$$f = \frac{(4)\left(960 \frac{\text{rev}}{\text{min}}\right)}{120 \frac{\text{s}}{\text{min}}} = 32 \text{ Hz}$$

This is not close to anything in commercial use. Try $p = 6$.

$$f = \frac{(6)\left(960 \frac{\text{rev}}{\text{min}}\right)}{120 \frac{\text{s}}{\text{min}}} = 48 \text{ Hz}$$

With a 4% slip, $f = 50$ Hz.

$$\boxed{50 \text{ Hz (European)}}$$

7. The required volt-amps per phase are determined by

$$\frac{Q}{3 \text{ phases}} = V_a I_a = \frac{V_a(E_g - V_a)}{X_s}$$
$$E_g = \left(\frac{Q}{3 \text{ phases}}\right)\left(\frac{X_s}{V_a}\right) + V_a$$
$$= \left(\frac{10^4 \text{ VA}}{3 \text{ phases}}\right)\left(\frac{20 \text{ }\Omega}{254 \text{ V}}\right) + \frac{440 \text{ V}}{\sqrt{3}}$$
$$= 516 \text{ V}$$
$$I_f = (516 \text{ V})\left(\frac{5 \text{ A}}{254 \text{ V}}\right)$$
$$I_r = E_g\left(\frac{I_f}{V_a}\right)$$
$$= \boxed{10.16 \text{ A}}$$

8.
$$P_{\text{in}} = \frac{P_{\text{out}}}{\text{efficiency}}$$
$$= \frac{(45 \text{ hp})\left(745.7 \frac{\text{W}}{\text{hp}}\right)}{0.93} = 36.1 \times 10^3 \text{ W}$$

At the minimum line current, the current is in phase with the voltage.

$$I = \frac{\dfrac{36.1 \times 10^3 \text{ W}}{3 \text{ phases}}}{\dfrac{440 \text{ V}}{\sqrt{3}}} = 47.37 \text{ A}$$

$$Z_{\text{base}} = \frac{254 \text{ V}}{47.37 \text{ A}} = 5.36 \text{ }\Omega$$

A synchronous reactance of 100% is common and is assumed.

$$X_s = 5.36 \ \Omega$$

$$I_a X_s = (5.36 \ \Omega)(47.37 \ \text{A}) = 254 \ \text{V}$$

The emf is

$$E_g = V_a - jX_s I_a = (254 \ \text{V})(1-j)$$

$$\left| E_g \right| = 359 \ \text{V}$$

The rotor current is given as $I_r = 8.2$ A.

To provide a leading power factor of 0.8, it is necessary to retain the same power.

$$|I_a|\cos\theta = 47.37 \ \text{A}$$

$$|I_a| = \frac{47.37 \ \text{A}}{0.8} = 59.2 \ \text{A}$$

$$I_a = (59.2 \ \text{A})(\cos\theta + j\sin\theta)$$

$$= 47.37 \ \text{A} + j35.52 \ \text{A}$$

$$= 59.2 \ \text{A}\angle 36.87°$$

$$E_g = 254 \ \text{V} - (j5.36 \ \Omega)I_a$$

$$= 512 \ \text{V}\angle{-29.74°}$$

Using a ratio gives

$$I_r = \left(\frac{8.2 \ \text{A}}{359 \ \text{V}}\right)(512 \ \text{V})$$

$$= \boxed{11.7 \ \text{A}}$$

9. The bases for the motor and transformer are the same.

$$I_{\text{rated}} = \frac{\dfrac{50 \times 10^3 \ \text{VA}}{3 \ \text{phases}}}{254 \ \text{V}} = 65.62 \ \text{A}$$

$$Z_{\text{base}} = \frac{254 \ \text{V}}{65.62 \ \text{A}} = 3.87 \ \Omega$$

From the blocked rotor test,

$$P + jQ = I^2(R_e + jX_e)$$

$$1097 \ \text{W} + j2791 \ \text{VAR} = (65.62 \ \text{A})^2(R_e + jX_e)$$

$$R_e + jX_e = 0.2548 + j0.6482 \ \Omega$$

On a per-unit basis,

$$R_e + jX_e = 0.0658 + j0.1675 \ \text{pu}$$

The per-unit circuit for starting is

By voltage division,

$$|V_{\text{line}}| = \left|\frac{0.0658 + j0.1675 \ \Omega}{0.0658 + j0.2675 \ \Omega}\right| = 0.653 \ \text{pu}$$

$$\text{dip} = \boxed{1 - 0.653 \ \text{pu} \ (35\%)}$$

10. The impedance for wye starting is three times that for delta starting, so the current will be only one-third the delta current.

$$I_{\text{starting}} = \frac{367 \ \text{A}}{3} = \boxed{122.3 \ \text{A}}$$

11. The synchronous speed is given by the following equation.

$$n_{\text{sync}} = \frac{120f}{p}$$

$$= \frac{(120)(400 \ \text{Hz})}{2 \ \text{poles}}$$

$$= 24{,}000 \ \text{rpm}$$

The answer is (D).

43 Electronics Fundamentals

PRACTICE PROBLEMS

1. Intrinsic germanium has a concentration of electrons of 2.5×10^{13} cm^{-3}. Determine the number of holes.

(A) 1.2×10^{13} cm^{-3}

(B) 1.5×10^{13} cm^{-3}

(C) 2.0×10^{13} cm^{-3}

(D) 2.5×10^{13} cm^{-3}

2. What type of semiconductor material results if phosphor is the only impurity?

(A) acceptor

(B) donor

(C) n-type

(D) p-type

3. A silicon semiconductor is doped with 10^{15} atoms cm^{-3} of phosphorus. What is the concentration of electrons?

(A) 1.6×10^{-5} cm^{-3}

(B) 2.6×10^{5} cm^{-3}

(C) 0.5×10^{15} cm^{-3}

(D) 1.0×10^{15} cm^{-3}

4. A silicon semiconductor is doped with a concentration of 10^{15} atoms cm^{-3} of phosphorus. What is the concentration of holes?

(A) 1.6×10^{-5} cm^{-3}

(B) 2.6×10^{5} cm^{-3}

(C) 0.5×10^{15} cm^{-3}

(D) 1.0×10^{15} cm^{-3}

5. A germanium semiconductor is doped with an impurity having three valence electrons. What are the majority carriers?

(A) conduction band electrons

(B) conduction band holes

(C) valence band electrons

(D) valence band holes

6. A discrete silicon semiconductor diode is forward biased at 0.75 V. What is the approximate current?

(A) 2.0 mA

(B) 3.0 nA

(C) 1.5 A

(D) 3.0 MA

7. The compound GaP, with an energy gap of 2.26 eV, is to be used in an LED. What is the expected wavelength of the emitted photons?

(A) 8.79×10^{-26} m

(B) 5.49×10^{-7} m

(C) 3.42×10^{12} m

(D) 5.46×10^{14} m

8. At 20°C, a germanium diode passes 70 μA when the reverse bias is −1.5 V. (a) What is the saturation current? (b) What is the current that flows when a forward bias of +0.2 V is applied at 20°C, and (c) at 40°C?

9. A voltage of $20 + 5\sqrt{2} \sin 60t$ V is applied to the circuit shown. The diode characteristics are a static forward resistance, r_f, of 120 Ω and a dynamic resistance, r_p, of 100 Ω. What is the voltage across the inductance?

10. At 25°C, a germanium diode shows a saturation current of 100 μA. What current is expected at 100°C, when the diode becomes "useless"?

11. At 25°C, a germanium diode shows a saturation current of 50 μA. What current is expected at 0°C?

12. At 25°C, a germanium diode shows a saturation current of 80 μA. At 25°C, what current is predicted for a voltage of −0.5 V?

Electronics

SOLUTIONS

1. In intrinsic semiconductors, $n = p$.

$$n_i^2 = np = n^2$$

$$n_i = \boxed{2.5 \times 10^{13} \text{ cm}^{-3}}$$

The answer is (D).

2. "Acceptor" and "donor" describe types of impurities. Phosphorus is from group VA on the periodic table, with five valence electrons. Therefore, it contributes a free electron. The semiconductor would be n-type.

The answer is (C).

3. Space-charge neutrality gives

$$N_A + n = N_D + p$$

Since $N_A = 0$ and $n \gg p$,

$$n \approx N_D = \boxed{1.0 \times 10^{15} \text{ cm}^{-3}}$$

The answer is (D).

4. From Prob. 3, $N_D \approx n$. The law of mass action becomes

$$n_i^2 = np = N_D p$$

$$p = \frac{n_i^2}{N_D} = \frac{(1.6 \times 10^{10} \text{ cm}^{-3})^2}{10^{15} \text{ cm}^{-3}}$$

$$= \boxed{2.56 \times 10^5 \text{ cm}^{-3} \quad (2.6 \times 10^5 \text{ cm}^{-3})}$$

The answer is (B).

5. Germanium has four valence electrons. The impurity has three valence electrons and thus adds a hole. The majority carriers will be holes.

The answer is (D).

6. The diode current is

$$I = I_s \left(e^{V/\eta V_T} - 1 \right)$$

A reverse saturation current of 10^{-9} A is typical for silicon, $\eta = 2$, and V_T at room temperature is

$$V_T = \frac{\kappa T}{q} = \frac{\left(1.381 \times 10^{-23} \frac{\text{J}}{\text{K}}\right)(300\text{K})}{1.6 \times 10^{-19} \text{ C}}$$

$$= 0.026 \text{ V}$$

Substituting the known and calculated values gives the diode current.

$$I = (10^{-9} \text{ A})\left(e^{0.75 \text{ V}/(2)(0.026 \text{ V})} - 1 \right)$$

$$= \boxed{1.84 \times 10^{-3} \text{ A} \quad (2.0 \text{ mA})}$$

The answer is (A).

7. The energy gap for GaP is 2.26 eV. The wavelength is

$$\lambda = \frac{hc}{E_G}$$

$$= \frac{(6.626 \times 10^{-34} \text{ J·s})\left(3.00 \times 10^8 \frac{\text{m}}{\text{s}}\right)}{(2.26 \text{ eV})\left(\frac{1.602 \times 10^{-19} \text{ J}}{1 \text{ eV}}\right)}$$

$$= \boxed{5.49 \times 10^{-7} \text{ m}}$$

The wavelength is in the visible spectrum.

The answer is (B).

8. (a)

$$I_s = \frac{I}{e^{qV/\eta \kappa T} - 1}$$

For a germanium diode, $\eta \approx 1$. At 20°C, $T_K = T_{°C} + 273° = 293$K.

$$I_s = \frac{70 \times 10^{-6} \text{ A}}{e^{\left(\frac{(1.6 \times 10^{-19} \text{ C})(-1.5 \text{ V})}{(1)\left(\frac{1.38 \times 10^{-23} \text{ J}}{\text{K}}\right)(293\text{K})}\right)} - 1}$$

$$\approx \boxed{\frac{70 \times 10^{-6} \text{ A}}{-1} \quad \left(-70 \ \mu\text{A}\right)}$$

The reverse saturation current is constant for reverse voltages up to the breakdown voltage.

(b)
$$I \approx I_s \left(e^{40 \text{ V}/\eta} - 1 \right)$$

$$= (70 \times 10^{-6} \text{ A})\left(e^{(40 \text{ V}^{-1})(0.2 \text{ V})/1} - 1 \right)$$

$$= \boxed{0.209 \text{ A}}$$

(c)
$$\frac{I_{s,40°C}}{I_{s,20°C}} = (2)^{(40°C - 20°C)/10°C}$$

$$(2)^2 = 4$$

$$I_{s,40°C} = 4 I_{s,20°C} = (4)(70 \times 10^{-6} \text{ A})$$

$$= 280 \times 10^{-6} \text{ A}$$

$$I = I_s \left(e^{qV/\eta \kappa T} - 1 \right)$$

$$= (280 \times 10^{-6} \text{ A})$$

$$\times \left(e^{\left(\frac{(1.6 \times 10^{-19} \text{ C})(0.2 \text{ V})}{(1)\left(\frac{1.38 \times 10^{-23} \text{ J}}{\text{K}}\right)(313\text{K})}\right)} - 1 \right)$$

$$= \boxed{0.4617 \text{ A}}$$

9. Because $20 > 5\sqrt{2}$, the applied voltage is never negative. Therefore, the diode does not rectify the applied voltage and serves no purpose. Model the circuit as an ideal diode with a 100 Ω resistance. (The forward resistance could also be used.)

$$X_L = \omega L = \left(60 \ \frac{\text{rad}}{\text{s}}\right)(1 \text{ H}) = 60 \ \Omega$$

$$Z = \sqrt{R^2 + X_L^2} = \sqrt{(100 \ \Omega)^2 + (60 \ \Omega)^2}$$

$$= 116.6 \ \Omega$$

$$\phi = \arctan \frac{X_L}{R} = \arctan \frac{60 \ \Omega}{100 \ \Omega} = 30.96°$$

$$\mathbf{I} = \frac{\mathbf{V}}{\mathbf{Z}} = \frac{20 \text{ V} + 5\sqrt{2} \text{ V}\angle 0°}{116.6 \ \Omega\angle 30.96°}$$

The DC voltage across the inductor is zero. Therefore, the 20 V bias is not used.

$$\mathbf{V}_L = \mathbf{I}\mathbf{Z}_L = \left(\frac{5\sqrt{2} \text{ V}\angle 0°}{116.6 \ \Omega\angle 30.96°}\right)(60 \ \Omega\angle 90°)$$

$$= \boxed{3.639 \text{ V}\angle 59.04°}$$

10. The saturation current doubles for every 10°C.

$$\frac{I_{s2}}{I_{s1}} = (2)^{\Delta T/10°C} = (2)^{(100°C - 25°C)/10°C}$$

$$= (2)^{7.5} = 181$$

$$I_{s2} = (181)(100 \times 10^{-6} \text{ A})$$

$$= \boxed{18,100 \times 10^{-6} \text{ A} \quad (18.1 \text{ mA})}$$

11. At 0°C,

$$I_{s2} = (2)^{(0°C - 25°C)/10°C}I_{s1} = (0.177)(50 \times 10^{-6} \text{ A})$$

$$= \boxed{8.85 \times 10^{-6} \text{ A} \quad (8.85 \ \mu\text{A})}$$

12. $\quad I_{-0.5} \approx I_s\left(e^{40V} - 1\right)$

$$= (80 \times 10^{-6} \text{ A})\left(e^{(40 \text{ V}^{-1})(-0.5 \text{ V})} - 1\right)$$

$$= \boxed{-80 \times 10^{-6} \text{ A} \quad (-80 \ \mu\text{A})}$$

44 Junction Transistors

PRACTICE PROBLEMS

1. Consider the common base (CB) circuit shown.

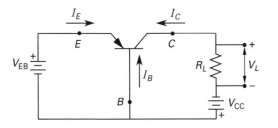

If the collector supply voltage has a magnitude of 1 V and the load is 50 Ω, determine and draw the load line.

2. A silicon *npn* transistor is used in the circuit shown. The transistor parameters are $I_{CBO} = 10^{-6}$ A and $\alpha_0 = 0.99$. The base-emitter voltage is 0.6 V. What is the quiescent collector current, I_C?

3. A silicon *npn* transistor is shown. Use $\alpha = 0.95$, $I_{CO} = 10^{-6}$ A, and $V_{BE} = 0.6$ V. What current, I_C, will result from this arrangement?

4. The equivalent circuit *h*-parameters for a bipolar junction transistor operating near the center of its active region in a common emitter configuration are $h_{ie} = 500$ Ω and $h_{fe} = 100$. Determine the value of h_{ib}.

5. The equivalent circuit *h*-parameters for a bipolar junction transistor operating near the center of its active region in a common emitter configuration are $h_{ie} = 500$ Ω and $h_{fe} = 50$. Determine the value of h_{fc}.

6. The equivalent circuit *h*-parameters for a bipolar junction transistor operating near the center of its active region in a common emitter configuration are $h_{ie} = 600$ Ω and $h_{fe} = 100$. Determine the value of h_{ic}.

7. The equivalent circuit *h*-parameters for a bipolar junction transistor operating near the center of its active region in a common emitter configuration are $h_{ie} = 500$ Ω and $h_{fe} = 200$. Determine the value of h_{fb}.

8. The equivalent circuit common base *h*-parameters for a bipolar junction transistor are $h_{ib} = 21.5$ Ω, $h_{fb} = -0.95$, $h_{rb} = 0.0031$, and $h_{ob} = 4.7 \times 10^{-7}$ $1/\Omega$. Determine the value of h_{ie}.

9. The *h*-parameters in the transistor depicted are $h_{ie} = 2200$ Ω, $h_{re} = 3.6 \times 10^{-4}$, $h_{fe} = 55$, and $h_{oe} = 12.5$ μS. (a) What load resistance is needed for a signal output voltage of 0.1 V with an input signal current of 0.5 μA? (b) What input voltage is required?

10. Consider the following figure of the characteristics of a bipolar junction transistor (BJT) configured as a common emitter.

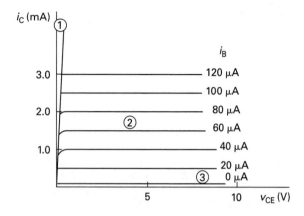

The bipolar junction transistor (BJT) whose characteristics are shown is to be used in a TTL (transistor-transistor logic) circuit. What operating region(s) will be utilized?

(A) 1

(B) 2

(C) 3

(D) 1 and 3

SOLUTIONS

1. Redraw the circuit as shown.

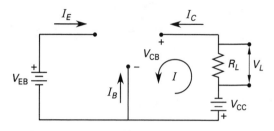

Kirchoff's voltage law (KVL) around the indicated loop gives

$$-V_{CC} - I_C R_L - V_{CB} = 0 \text{ V}$$

Because $I_C = 0$ A and $V_{CC} = -1$ V,

$$V_{CC} = V_{CB} = -1 \text{ V}$$

Redraw the current as shown.

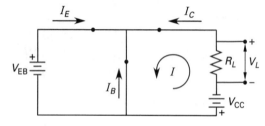

Using Ohm's law and the given information,

$$I_C = \frac{V_{CC}}{R_L} = \frac{-1 \text{ V}}{50 \ \Omega} = -20 \times 10^{-3} \text{ A} \quad (-20 \text{ mA})$$

Plot V_{CC} and I_C, and connect them with a straight line. The result is normally plotted as shown.

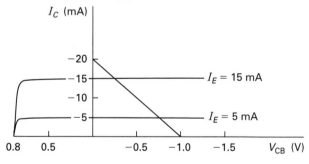

2.
$$I_C = \alpha I_E - I_{CBO}$$

$$= \alpha \left(\frac{V_{BB} - V_{BE}}{R_E} \right) - I_{CBO}$$

$$= (0.99) \left(\frac{4 \text{ V} - 0.6 \text{ V}}{1000 \ \Omega} \right) - 10^{-6} \text{ A}$$

$$= \boxed{3.365 \times 10^{-3} \text{ A} \quad (3.365 \text{ mA})}$$

Alternate Solution

Writing KVL around the base loop,

$$V_{BB} - V_{BE} - (I_C + I_B)R_E = 0$$

$$\frac{I_C}{I_B} = \frac{\alpha}{1-\alpha} = \frac{0.99}{1-0.99} = 99$$

$$4\text{ V} - V_{BE} - \left(I_C + \frac{I_C}{99}\right)(1000\ \Omega) = 0$$

$$4\text{ V} - 0.6\text{ V} - (1010\ \Omega)I_C = 0$$

$$I_C = \boxed{3.366 \times 10^{-3}\text{ A }(3.366\text{ mA})}$$

3. R_1 and R_2 form a voltage divider.

$$V_B = \left(\frac{R_2}{R_1 + R_2}\right)V_{CC}$$

$$= \left(\frac{60\text{ k}\Omega}{30\text{ k}\Omega + 60\text{ k}\Omega}\right)(24\text{ V})$$

$$= 16\text{ V}$$

For the purpose of the signal, R_1 and R_2 both connect to ground and are in parallel.

$$R_{in} = \frac{R_1 R_2}{R_1 + R_2} = \frac{(30\text{ k}\Omega)(60\text{ k}\Omega)}{30\text{ k}\Omega + 60\text{ k}\Omega}$$

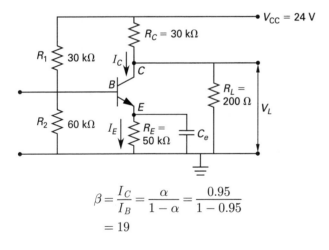

$$\beta = \frac{I_C}{I_B} = \frac{\alpha}{1-\alpha} = \frac{0.95}{1-0.95}$$

$$= 19$$

So, $I_B = I_C/19$.

Writing KVL around the base circuit,

$$V_B - V_{BE} - (I_C + I_B)R_E - I_B R_{in} = 0\text{ V}$$

$$16\text{ V} - 0.6\text{ V} - \left(I_C + \frac{I_C}{19}\right)(50\text{ k}\Omega)\left(1000\ \frac{\Omega}{\text{k}\Omega}\right)$$

$$- \left(\frac{I_C}{19}\right)(20\text{ k}\Omega)\left(1000\ \frac{\Omega}{\text{k}\Omega}\right) = 0\text{ V}$$

$$I_C = 2.869 \times 10^{-4}\text{ A }(0.2869\text{ mA})$$

I_{CO} is given, so it should be used even though its effect is small.

$$I'_C = I_C + I_{CO} = 2.869 \times 10^{-4}\text{ A} + 10^{-6}\text{ A}$$

$$= \boxed{2.879 \times 10^{-4}\text{ A }\quad(0.2879\text{ mA})}$$

The exact solutions, based on node-voltage analysis, are

$$V_B = 15.70\text{ V}$$

$$I_B = 0.0151\text{ mA}$$

$$I_C = 0.2869\text{ mA}$$

4.
$$h_{ib} = \frac{h_{ie}}{1 + h_{fe}} = \frac{500\ \Omega}{1 + 100}$$

$$= \boxed{4.95\ \Omega}$$

5.
$$h_{fc} = -1 - h_{fe} = -1 - 50$$

$$= \boxed{-51}$$

6.
$$h_{ic} = h_{ie}$$

$$= \boxed{600\ \Omega}$$

7.
$$h_{fb} = \frac{-h_{fe}}{1 + h_{fe}} = \frac{-200}{1 + 200}$$

$$= \boxed{-1}$$

8.
$$h_{ie} = \frac{h_{ib}}{1 + h_{fb}} = \frac{21.5\ \Omega}{1 + (-0.95)}$$

$$= \boxed{430\ \Omega}$$

9.

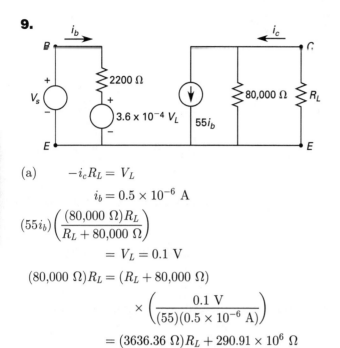

(a)
$$-i_c R_L = V_L$$
$$i_b = 0.5 \times 10^{-6} \text{ A}$$
$$(55 i_b)\left(\frac{(80{,}000 \ \Omega) R_L}{R_L + 80{,}000 \ \Omega}\right)$$
$$= V_L = 0.1 \text{ V}$$
$$(80{,}000 \ \Omega) R_L = (R_L + 80{,}000 \ \Omega)$$
$$\times \left(\frac{0.1 \text{ V}}{(55)(0.5 \times 10^{-6} \text{ A})}\right)$$
$$= (3636.36 \ \Omega) R_L + 290.91 \times 10^6 \ \Omega$$
$$R_L = \boxed{3809 \ \Omega}$$

(b)
$$V_s = i_b(2200 \ \Omega) + (3.6 \times 10^{-4}) V_L$$
$$- (0.5 \times 10^{-6} \text{ A})(2200 \ \Omega)$$
$$+ (3.6 \times 10^{-4})(0.1 \text{ V})$$
$$= \boxed{1.136 \times 10^{-3} \text{ V} \quad (1.136 \text{ mV})}$$

10. Logic circuits require the transistor to operate as a switch, thereby providing two outputs (logic 0 and logic 1). Logic circuit operation requires the transistor to operate between the saturation region (1) and the cutoff region (3).

The answer is (D).

45 Field Effect Transistors

PRACTICE PROBLEMS

1. An AC-coupled FET amplifier circuit is shown.

Assume the drain bias voltage is $V_{DD} = 20$ V. If I_D at $V_{DS} = 0$ V is 10 mA, draw the load line.

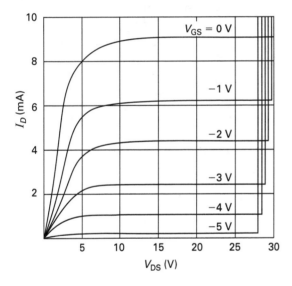

2. A FET with the characteristics shown is to be used in the circuit in Prob. 1. $R_D = 1500$ Ω, $V_{DD} = 20$ V, and $V_{GSQ} = -2$ V. Draw the DC load line and locate the Q-point.

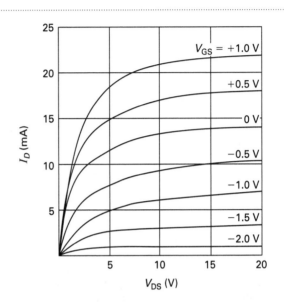

3. The manufacturer of a MOSFET has provided the following information: $V_{DS,max} = 20$ V, $V_{GS,min} = -8$ V, $I_{D,max} = 25$ mA, and $P_{max} = 330$ mW. The FET is to be used in the amplifier circuit shown, with $R_D = 800$ Ω and $R_L = 800$ Ω. Draw the DC load line and a Q-point to ensure linear amplification. Use the characteristic curves from Prob. 2.

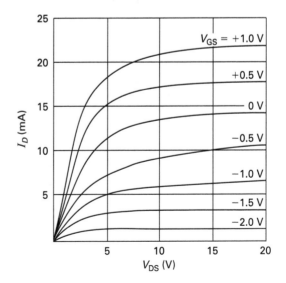

4. A voltage source produces a triangular wave shape as shown.

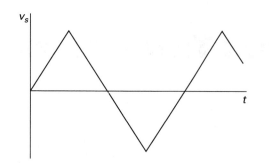

An output wave shape of the indicated form is desired.

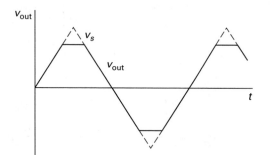

Consider the four circuits shown. Which of the indicated circuits can provide the desired output?

(A)

(B)

(C)

(D)

5. Consider the following figure of a MOSFET.

What type of device is the MOSFET?

    (A) n-channel enhancement

    (B) n-channel depletion

    (C) p-channel enhancement

    (D) p-channel depletion

SOLUTIONS

1.

$$P_1(V_{DS} = 20 \text{ V}, \ I_D = 0 \text{ A})$$
$$P_2(V_{DS} = 0 \text{ V}, \ I_D = 10 \text{ mA})$$

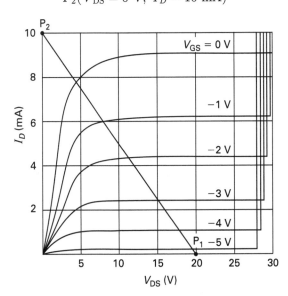

2.

$$P_1(V_{DS} = 20 \text{ V}, 0)$$
$$P_2\left(0, I_D = \frac{20 \text{ V}}{1500 \ \Omega} = 0.01333 \text{ A} \quad (13.33 \text{ mA})\right)$$

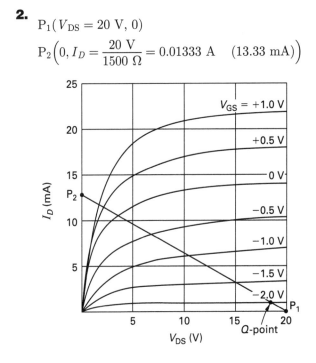

3. The highly negative V_{GS} curves are not shown and are, presumably, crowded together near the $I_D = 0$ A axis. Because of this, if the Q-point were in the highly negative region, positive swings in V_{GS} would accompany large increases in I_D, but negative swings in V_{GS} would decrease I_D very little. Therefore, the quiescent operating point needs to be near the $V_{GS} = 0$ V line. In this position, changes in I_D and V_{GS} are approximately proportional. Power dissipation is another consideration. It should be evaluated at the extreme points of the signal swing, not only at the Q-point.

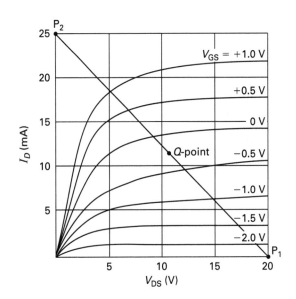

4. A clipping (limiting) circuit that clamps the voltage on both the positive and negative portion of the cycles is needed. Therefore, only choices (A) and (D) are possibilities.

Consider both the figures for choices (A) and (D). During the positive half cycle, diode D^+ (associated with V_{ref+}) conducts once the source voltage is greater than the reference voltage. This means that in a portion of the positive half cycle, the diode V_{ref+} conducts, limiting the voltage drop to the reference value. Diode D^- (associated with V_{ref-}) is reverse biased and does not conduct.

Consider the negative half cycle. In choice (A), diode D^- conducts almost immediately, thereby clamping the output voltage to near zero. In choice (D), diode D^- conducts once the source voltage is less than the reference voltage. Or, one can say that diode D^- conducts when it overcomes V_{ref-}. (A symmetrical wave is generated when $|V_{ref+}| = |V_{ref-}|$.)

The answer is (D).

5. The dashed line on the MOSFET indicates an *enhancement* device. Some symbology uses a single connected line for an enhancement device and a thicker line for a depletion device. The thicker line is meant to indicate that a channel is already present. The alternate symbology is shown.

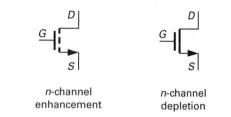

n-channel enhancement   *n*-channel depletion

The arrow indicates the direction of conventional current flow. Therefore, the current flows from the drain to the source. That is, positive charges flow from the drain to the source, meaning that negative charges flow from the source (the source of the majority charges) to the drain. The channel is composed of these negative charges, so this is an *n*-channel enhancement device.

The answer is (A).

46 Electrical and Electronic Devices

PRACTICE PROBLEMS

1. An operational amplifier has a gain of 10^5. The frequencies at the half-power points are $f_L = 500$ Hz and $f_H = 1000$ Hz. What is the figure of merit?

(A) 1×10^6 rad/s

(B) 5×10^7 rad/s

(C) 3×10^8 rad/s

(D) 6×10^8 rad/s

2. During testing of an operational amplifier, a signal of equal magnitude is applied to the two terminals. That is, $v^+ = -v^-$. What gain is being measured?

(A) A_{cm}

(B) A_{dm}

(C) A_V

(D) G_P

3. An operational amplifier data sheet indicates a voltage gain of 10^5 and a power supply voltage of 12 V. What is the maximum voltage difference between the input terminals that will ensure linear operation?

(A) 60 μV

(B) 90 μV

(C) 120 μV

(D) 150 μV

4. Consider the circuit shown. The feedback resistance is 10 kΩ, and the input resistance is 100 Ω. If the op amp gain is 10^5, what is the gain of the circuit?

(A) -100

(B) -99.9

(C) $+100$

(D) 10^5

5. An operational amplifier has an input resistance of 10^5 Ω and a bandwidth of 1000 Hz, and is at a room temperature of 300K. If 10 dB above the noise is required for proper operation, what is the required signal power?

(A) 0.10 μV

(B) 2.0 μV

(C) 4.0 μV

(D) 16 μV

6. Consider the circuit shown. (a) If $R_f = 1$ MΩ and $R_{in} = 50$ Ω, what is the gain? (b) If $v_{in} = 1$ mV, what is v_{out}?

7. For the circuit shown, $v_1 = 10\sin 200t$ and $v_2 = 15\sin 200t$. What is v_{out}? The op amp is ideal and has infinite gain.

8. In the circuit shown, if v_{in} is $\sin 30t$, what is v_{out}? The op amp is ideal and has infinite gain.

9. In the circuit shown, if $v_1 = v_2 = v_3 = 12$ V, what is v_{out}? The op amp is ideal and has infinite gain.

10. Given the following circuit, (a) what value of R_f will give an output of -8 V? (b) If the maximum of each input is 10 V and the output must not exceed -15 V, what value must be used for R_f? The op amp is ideal and has infinite gain.

11. An integrator is used in a device for measuring displacement from a fixed location. The scale of the device is 1 V to 1 in. The switch is closed to set the output for initial displacement and then opened at $t=0$ s to initiate measurement. If the displacement is initially $d_0 = 12$ in, what is the output voltage for an input of $8\cos 10t$ V? The op amp is ideal and has infinite gain.

12. What differential equation is solved by the following circuit? The op amp is ideal and has infinite gain.

13. For the circuit shown, what is the differential equation that writes v_{out} in terms of v_{in}? The op amp is ideal and has infinite gain.

14. For the circuit shown, determine the output voltage as a function of the input voltage.

15. The op amp circuit shown is designed to provide a characteristic similar to a relay. When the output is negative and the input current increases, the output will go to the positive limit of +12 V when the input voltage exceeds +1 V. It will remain there until the input voltage becomes less than −1 V, at which time the output will go to the negative limit of −12 V. Determine an appropriate value for R.

16. A germanium diode with an offset voltage of 0.2 V is used in a precision diode circuit. The power supply voltages are ±15 V. The amplifier voltage gain is 10^7. What is the threshold voltage of the precision diode?

(A) 20 nV

(B) 100 nV

(C) 1.5 μV

(D) 9.0 μV

The following circuit is applicable to Prob. 17 through Prob. 20. The output is to be maintained at 24 V, while the input varies from 120 V to 126 V with a load current ranging from 0 A to 5 A. Assume ideal conditions, that is, the keep-alive current is zero and the zener resistance, R_Z, is negligible.

17. What is the required supply resistance to ensure proper operation of the zener regulator circuit?

(A) 19 Ω

(B) 20 Ω

(C) 24 Ω

(D) 25 Ω

18. What is the maximum zener current?

(A) 4.1 A

(B) 4.3 A

(C) 5.1 A

(D) 5.3 A

19. What is the minimum diode power rating that can be used?

(A) 100 W

(B) 150 W

(C) 300 W

(D) 600 W

20. What is the minimum power rating of the supply resistor?

(A) 400 W

(B) 450 W

(C) 500 W

(D) 550 W

The following circuit is applicable to Prob. 21 through Prob. 24. The circuit is to be operated at low frequencies. The transistor has the following h-parameter values: $h_{\mathrm{ie}} = 4 \times 10^3 \, \Omega$, $h_{\mathrm{fe}} = 200$, $h_{\mathrm{oe}} = 100 \times 10^{-6} \, \mathrm{S}$, and $h_{\mathrm{re}} = 4 \times 10^3$.

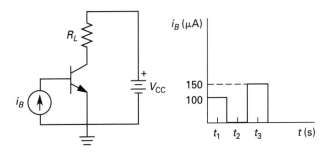

21. If the load resistance is 400 Ω, what collector supply voltage is required for the transistor to be on at time t_1?

(A) 4 V

(B) 8 V

(C) 12 V

(D) 16 V

22. If the collector supply voltage is 12 V and the load resistor is 400 Ω, what is the condition of the transistor at time t_2?

(A) active

(B) off

(C) on

(D) saturated

23. The selected collector supply voltage is 12 V. The transistor must be on for pulses of 150 μA or greater only. What is the required value of the load resistance?

(A) 200 Ω

(B) 300 Ω

(C) 400 Ω

(D) 600 Ω

24. For the circuit parameters in Prob. 23, what is the actual saturation current for a silicon transistor?

(A) 20.0 mA

(B) 25.5 mA

(C) 29.5 mA

(D) 30.0 mA

25. Consider the inverter, or NOT gate, circuit shown.

Parameters taken from a manufacturer's data sheet indicate that $V_{CC} = 5.0$ V, $I_{B,sat} > 0.5$ mA, $V_{BE,sat} < 1.5$ V, $R_c = 500$ Ω, and $R_b = 1000$ Ω. By how much does a reverse saturation current of 1 mA lower the fan-out of $Q1$?

(A) 2

(B) 4

(C) 10

(D) 12

26. For the transistor circuit shown, determine (a) the necessary base current for this gate to sink five identical gates connected to its output, and (b) the minimum size of R_b to allow this gate to source five identical gates.

27. A particular logic gate has the following parameters: $t_d = 5$ ns, $t_r = 50$ ns, $t_s = 30$ ns, and $t_f = 30$ ns. Determine the maximum frequency at which this gate can operate.

28. Use an operational amplifier to convert a rectangular wave voltage of ± 5 V and a period of 10 ms to a triangular wave with a peak output of ±5 V with the same period.

29. For the circuit shown, determine the transfer characteristic of the output voltage versus the input voltage. The diodes are ideal.

30. Determine the power requirement for the DTL circuit shown for a fan-out of five.

SOLUTIONS

1. The figure of merit is

$$F_m = A_{\text{ref}}\,(\text{BW}) = A_{\text{ref}}\,(2\pi f_H - 2\pi f_L)$$
$$= (10^5)2\pi(1000\text{ Hz} - 500\text{ Hz})$$
$$= \boxed{3.14 \times 10^8\ \text{rad/s}\quad(3 \times 10^8\ \text{rad/s})}$$

The answer is (C).

2. The differential-mode voltage is

$$v_{\text{dm}} = v^+ - v^-$$

Therefore, with $v^+ = -v^-$,

$$v_{\text{dm}} = v^+ + v^+ = 2v^+$$

The common-mode voltage is

$$v_{\text{cm}} = \tfrac{1}{2}\left(v^+ + v^-\right) = \tfrac{1}{2}(v^+ + (-v^-)) = 0\ \text{V}$$

Therefore, only the difference is being amplified, and the gain so determined is A_{dm}.

The answer is (B).

3. The maximum voltage difference is

$$|v^+ - v^-| < \frac{V_{\text{DC}} - 3\text{ V}}{A_V}$$
$$|\Delta V| < \frac{12\text{ V} - 3\text{ V}}{10^5} = \frac{9\text{ V}}{10^5}$$
$$= \boxed{90 \times 10^{-6}\ \text{V}\quad(90\ \mu\text{V})}$$

The answer is (B).

4. Consider the circuit shown, assuming the indicated current directions at the negative terminal.

Electronics

Writing Kirchoff's current law (KCL) at the inverting terminal node gives

$$\frac{V^- - v_{in}}{R_{in}} + \frac{V^- - v_{out}}{R_f} + i^- = 0$$

Applying the ideal op amp assumptions means $i^- = 0$ A and $v^- = 0$ V. Substituting and solving for the gain gives

$$\frac{0 - v_{in}}{R_{in}} + \frac{0 - v_{out}}{R_f} + 0 = 0$$

$$\frac{v_{out}}{v_{in}} = \frac{-R_f}{R_{in}} = A_{V,\text{ideal}}$$

$$= \frac{-10 \times 10^3\ \Omega}{100\ \Omega} = -100$$

But this op amp is real and has a finite gain. The constraint equation (that is, the fundamental op amp equation) rearranged is

$$V^- - V^+ = \Delta = \frac{-v_{out}}{A_V}$$

Since V^+ is at 0 V,

$$V^- = \frac{-v_{out}}{A_V}$$

Substituting into the original equation gives

$$\frac{\dfrac{-v_{out}}{A_V} - v_{in}}{R_{in}} + \frac{\dfrac{-v_{out}}{A_V} - v_{out}}{R_f} + 0 = 0$$

The input current is still small enough to be ignored. Rearranging gives

$$\frac{v_{out}}{v_{in}} = -A_V\left(\frac{R_f}{R_f + R_{in} + A_V R_{in}}\right)$$

When the gain is large, this expression reduces to the expression for gain in the ideal case.

Substituting the given values gives the following result.

$$\frac{v_{out}}{v_{in}} = (-10^5)\left(\frac{10 \times 10^3\ \Omega}{10 \times 10^3\ \Omega + 100\ \Omega + (10^5)(100\ \Omega)}\right)$$

$$= \boxed{-99.9}$$

The answer is (B).

5. The signal power is found from the signal to noise ratio (SNR).

$$\text{SNR} = 10 \log \frac{P_s}{P_n}$$

$$\text{antilog}\,\frac{\text{SNR}}{10} = \frac{P_s}{P_n}$$

$$P_s = P_n\,\text{antilog}\,\frac{\text{SNR}}{10}$$

The noise power is unknown. The noise power is given in terms of the Boltzmann constant, k; temperature, T, in kelvins; and the bandwidth, BW, in hertz.

$$P_n = \frac{V_{rms}^2}{4R} = (\text{BW})kT$$

$$= (1000\ \text{Hz})\left(1.3805 \times 10^{-23}\ \frac{\text{J}}{\text{K}}\right)(300\text{K})$$

$$= 4.14 \times 10^{-18}\ \text{W}$$

Substituting,

$$P_s = (4.14 \times 10^{-18}\ \text{W})\,\text{antilog}\,\frac{10}{10}$$

$$= 41.4 \times 10^{-18}\ \text{W}$$

The voltage is found from

$$P_s = \frac{V_{rms}^2}{4R}$$

$$V_{rms} = \sqrt{4RP_s} = \sqrt{(4)(10^5\ \Omega)(41.4 \times 10^{-18}\ \text{W})}$$

$$= \boxed{4.07 \times 10^{-6}\ \text{V} \quad (4.0\ \mu\text{V})}$$

The answer is (C).

6. (a) The gain is

$$A = \frac{-R_f}{R_{in}} = \frac{-10^6\ \Omega}{50\ \Omega} = \boxed{-20{,}000}$$

(b) $v_{out} = Av_{in} = (-20{,}000)(0.001\ \text{V}) = \boxed{-20\ \text{V}}$

7. This is a summing circuit.

$$v_{\text{out}} = -\left(\frac{Z_f}{Z_1}\right)v_1 - \left(\frac{Z_f}{Z_2}\right)v_2$$

$$= -\left(\frac{1}{sC}\right)\left(\frac{v_1}{R_1} + \frac{v_2}{R_2}\right)$$

$$= \frac{-1}{(2 \times 10^{-6}\ \text{F})(0.75 \times 10^6\ \Omega)}\int 10\sin 200t\, dt$$

$$\qquad - \frac{1}{(2 \times 10^{-6}\ \text{F})(0.5 \times 10^6\ \Omega)}\int 15\sin 200t\, dt$$

$$= \left(\frac{2}{3}\right)\left(\frac{1}{200}\right)(10)\cos 200t$$

$$\qquad + \left(\frac{1}{1}\right)\left(\frac{15}{200}\right)\cos 200t + C$$

$$= \boxed{\left(\frac{1}{30}\right)\cos 200t + \left(\frac{3}{40}\right)\cos 200t + C\ \text{V}}$$

8. The circuit is a differentiator.

$$v_{\text{out}} = -RC\frac{dv_{\text{in}}}{dt}$$

$$= -(0.8 \times 10^6\ \Omega)(2 \times 10^{-6}\ \text{F})\left(\frac{d(\sin 30t)}{dt}\right)$$

$$= \boxed{-48\cos 30t\ \text{V}}$$

9. This is a summing circuit.

$$v_{\text{out}} = \left(\frac{-6\ \text{k}\Omega}{4.5\ \text{k}\Omega}\right)v_1 - \left(\frac{6\ \text{k}\Omega}{5\ \text{k}\Omega}\right)v_2$$

$$\qquad - \left(\frac{6\ \text{k}\Omega}{7.5\ \text{k}\Omega}\right)v_3$$

$$= (-12\ \text{V})\left(\begin{array}{c}\dfrac{6\ \text{k}\Omega}{4.5\ \text{k}\Omega} + \dfrac{6\ \text{k}\Omega}{5\ \text{k}\Omega} \\[2mm] + \dfrac{6\ \text{k}\Omega}{7.5\ \text{k}\Omega}\end{array}\right)$$

$$= \boxed{-40\ \text{V}}$$

10. (a) This is a summing circuit.

$$v_{\text{out}} = -\left(\begin{array}{c}\left(\dfrac{R_f}{1\ \Omega}\right)(1\ \text{V}) + \left(\dfrac{R_f}{1\ \Omega}\right)(6\ \text{V}) \\[2mm] + \left(\dfrac{R_f}{1\ \Omega}\right)(3\ \text{V})\end{array}\right) = -8\ \text{V}$$

$$R_f = \boxed{0.8\ \Omega}$$

(b)

$$v_{\text{out}} = -\left(\begin{array}{c}\left(\dfrac{R_f}{1\ \Omega}\right)(10\ \text{V}) + \left(\dfrac{R_f}{1\ \Omega}\right)(10\ \text{V}) \\[2mm] + \left(\dfrac{R_f}{1\ \Omega}\right)(10\ \text{V})\end{array}\right) = -15\ \text{V}$$

$$R_f = \boxed{0.5\ \Omega}$$

11. $v_{\text{out}} = -\dfrac{1}{RC}\displaystyle\int v_{\text{in}}\, dt$

$$= -\left(\begin{array}{c}\dfrac{1}{(500{,}000\ \Omega)} \\[2mm] \times (2 \times 10^{-3}\ \text{F})\end{array}\right)\int 8\cos 10t\, dt$$

$$= (-8 \times 10^{-4})\sin 10t + V_0$$

At $t = 0$ s, $d_0 = 12$ in.

$$V_0 = (12\ \text{in})\left(\frac{1\ \text{V}}{1\ \text{in}}\right) = 12\ \text{V}$$

$$v_{\text{out}} = \boxed{(-8 \times 10^{-4})\sin 10t + 12\ \text{V}}$$

12. $\dfrac{v_{\text{out}}}{v_{\text{in}}} = -\dfrac{Z_f}{Z_{\text{in}}} = \dfrac{-500{,}000\ \Omega}{\dfrac{(10^6\ \Omega)\left(\dfrac{1}{s(2 \times 10^{-6})}\ \Omega\right)}{10^6\ \Omega + \dfrac{1}{s(2 \times 10^{-6})}\ \Omega}}$

$$= -\frac{\dfrac{1}{2}}{\dfrac{1}{2s + 1}} = -\left(s + \tfrac{1}{2}\right)$$

$$v_{\text{out}} = -\left(s + \tfrac{1}{2}\right)v_{\text{in}}$$

$$= \boxed{-\frac{dv_{\text{in}}}{dt} - \frac{v_{\text{in}}}{2}}$$

13. $\dfrac{v_{\text{out}}}{v_{\text{in}}} = -\dfrac{Z_f}{Z_{\text{in}}}$

$$= -\frac{\dfrac{(600{,}000\ \Omega)\left(\dfrac{1}{s(2 \times 10^{-6})}\ \Omega\right)}{600{,}000\ \Omega + \left(\dfrac{1}{s(2 \times 10^{-6})}\ \Omega\right)}}{2{,}000{,}000\ \Omega}$$

$$= -\frac{0.3}{1.2s + 1}$$

$$v_{out}(1.2s + 1) = -0.3v_{in}$$

$$(1.2)\left(\frac{dv_{out}}{dt}\right) + v_{out} = -0.3v_{in}$$

$$\boxed{(1.2)\left(\frac{dv_{out}}{dt}\right) + v_{out} + 0.3v_{in} = 0}$$

14.

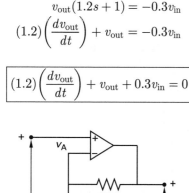

$$v_A = 0 \text{ V}$$

$$v_B = 0 \text{ V}$$

$$v_{in} = i(1 \text{ k}\Omega)$$

$$i = \frac{v_{in}}{1 \text{ k}\Omega}$$

$$v_{out} = (4 \text{ k}\Omega + 1 \text{ k}\Omega + 4 \text{ k}\Omega)i$$

$$= \left(\frac{9 \text{ k}\Omega}{1 \text{ k}\Omega}\right)v_{in}$$

$$= \boxed{9v_{in}}$$

15. Except while slewing from one saturation level to the other, the op amp is not operating in its linear region.

The current flowing in the resistors is

$$i_{in} = \frac{v_{in} - v_{out}}{1 \text{ k}\Omega + R}$$

The voltage of the positive terminal is

$$v_{in} - i_{in}(1 \text{ k}\Omega) = v_{in} - \frac{(1 \text{ k}\Omega)v_{in} - (1 \text{ k}\Omega)v_{out}}{1 \text{ k}\Omega + R}$$

$$v^+ = \frac{Rv_{in} + (1 \text{ k}\Omega)v_{out}}{R + 1 \text{ k}\Omega}$$

For v^+ positive, v_{out} is $+12$ V. v_{out} will reverse when v^+ becomes negative.

$$v_{in} = (-1 - \delta) \text{ V}$$

$$-R(1 \text{ V} + \delta) + (1 \text{ k}\Omega)(12 \text{ V}) < 0$$

$$-R + 12 \text{ k}\Omega - \delta R < 0$$

Let δ approach zero. Then,

$$R = \boxed{12 \text{ k}\Omega}$$

16. The threshold voltage is reduced by an amount equal to the second term in Eq. 46.17, which is the transfer equation for a precision diode. The forward threshold voltage drop is V_F, and A_V is the voltage gain.

$$V_{out} = V_{in} - \frac{V_F}{A_V}$$

Substituting the given information gives

$$\frac{V_F}{A_V} = \frac{0.2 \text{ V}}{10^7} = \boxed{20 \times 10^{-9} \text{ V} \quad (20 \text{ nV})}$$

The answer is (A).

17. For a zener regulator circuit, the supply resistance is given by Eq. 46.20. V_{ZM} is the voltage zener at maximum rated current. $I_{L,max}$ is the maximum load current.

$$R_s = \frac{V_{in,min} - V_{ZM}}{I_{L,max}}$$

$$= \frac{120 \text{ V} - 24 \text{ V}}{5 \text{ A}}$$

$$= \boxed{19.2 \ \Omega \quad (19 \ \Omega)}$$

The answer is (A).

18. The maximum zener current is determined from Eq. 46.21.

$$I_{Z,max} = \left(\frac{V_{in,max} - V_{ZM}}{V_{in,min} - V_{ZM}}\right) I_{L,max}$$

$$= \left(\frac{126 \text{ V} - 24 \text{ V}}{120 \text{ V} - 24 \text{ V}}\right)(5 \text{ A})$$

$$= \boxed{5.3 \text{ A}}$$

The answer is (D).

Electronics

19. The diode power is given by Eq. 46.22.

$$P_D = I_{Z,\text{max}} V_{\text{ZM}} = (5.3\text{ A})(24\text{ V})$$
$$= 127.2\text{ W}$$

The minimum diode power rating that can be used is 150 W.

The answer is (B).

20. The minimum power rating of the supply resistor is given by Eq. 46.23.

$$P_{R_s} = I_{Z,\text{max}}^2 R_s$$
$$= (5.3\text{ A})^2 (19.2\ \Omega)$$
$$= \boxed{539.3\text{ W} \quad (550\text{ W})}$$

The answer is (D).

21. At low frequencies h_{fe} is approximately equal to h_{FE}, so the small-signal parameter given can be used. The base current, I_B, necessary for saturation is given by Eq. 46.37. V_{CC} is the collector supply voltage. R_L is the load resistance.

$$I_B \geq \frac{V_{\text{CC}}}{h_{\text{FE}} R_L}$$

For a load resistance of 400 Ω, the collector supply voltage is

$$\begin{aligned} V_{\text{CC}} &\leq I_{B_{t1}} h_{\text{FE}} R_L \\ &\leq (100 \times 10^{-6}\text{ A})(200)(400\ \Omega) \\ &\leq \boxed{8\text{ V}} \end{aligned}$$

The answer is (B).

22. If the base current is zero, the transistor is off.

The answer is (B).

23. The minimum base current, which will be 150 μA to ensure that the transistor is not on at t_1, is given by Eq. 46.37.

$$I_B \geq \frac{V_{\text{CC}}}{h_{\text{FE}} R_L} \approx \frac{V_{\text{CC}}}{h_{\text{fe}} R_L}$$

Rearranging gives

$$\begin{aligned} R_L &\geq \frac{V_{\text{CC}}}{h_{\text{fe}} I_B} \\ &\geq \frac{12\text{ V}}{(200)(150\ \mu\text{A})\left(\dfrac{1\text{ A}}{1 \times 10^6\ \mu\text{A}}\right)} \\ &\geq \boxed{400\ \Omega} \end{aligned}$$

The answer is (C).

24. The actual saturation current is given by Eq. 46.36. $V_{\text{CE,sat}}$ is the collector-emitter voltage at saturation.

$$\begin{aligned} I_{C,\text{sat}} &= \frac{V_{\text{CC}} - V_{\text{CE,sat}}}{R_L} \\ &= \frac{12\text{ V} - 0.2\text{ V}}{400\ \Omega} \\ &= \boxed{29.5 \times 10^{-3}\text{ A} \quad (29.5\text{ mA})} \end{aligned}$$

If the collector-emitter voltage drop is ignored, the saturation current will be 30 mA. Except for exacting calculations, the ON transistor can be modeled as a short circuit.

The answer is (C).

25. A transistor inverter is a common-emitter configured transistor as shown.

Initially ignoring the reverse saturation current, I_{CBO}, means $Q2$ can be modeled as an open circuit. Then, writing Kirchoff's voltage law (KVL) from the reference ground at $+V_{\text{CC}}$ (not shown) back to the reference ground (shown at $Q2$), which comprises a loop, gives

$$V_{\text{CC}} - N I_{B,\text{min}} R_c - I_{B,\text{min}} R_b - V_{\text{BE2}} = 0$$

The minimum base current is the current necessary to cause $Q2$ and other load transistors to operate in the saturation region. This is given as $I_{B,sat} > 0.5$ mA. The base-emitter voltage is the voltage of the transistor while saturated, given as $V_{BE,sat} < 1.5$ V. Substituting and solving for the fan-out gives

$$N = \frac{V_{CC} - V_{BE,sat} - I_{B,min}R_b}{I_{B,min}R_c}$$

$$= \frac{5 \text{ V} - 1.5 \text{ V} - (0.5 \times 10^{-3} \text{ A})(1000 \ \Omega)}{(0.5 \times 10^{-3} \text{ A})(500 \ \Omega)}$$

$$= 12$$

To account for I_{CBO}, add the additional voltage drop $I_{CBO}R_c$ to the original KVL equation, or use Eq. 46.45 directly.

$$V_{CC} - NI_{B,min}R_c - I_{CBO}R_c - I_{B,min}R_b - V_{BE2} = 0$$

$$N + \frac{R_b}{R_c} = \frac{V_{CC} - V_{BE2} - I_{CBO}R_c}{I_{B,min}R_c}$$

Substituting and solving gives

$$N = 10$$

The reverse saturation current lowers the fan-out by

$$12 - 10 = \boxed{2}$$

The answer is (A).

26. The load line with no loads has $I_C \approx 4.9$ mA, which requires I_B to be slightly more than 0.3 mA.

(a) To sink five additional loads, because $I_{CBO} = 0.4$ mA, I_C must increase by $(5)(0.4 \text{ mA}) = 2$ mA to approximately 6.9 mA. This will require $I_B \approx 0.4$ mA.

$$I_{B,\text{low output}} = \boxed{0.4 \text{ mA}}$$

(b) When this transistor is off, the five loads draw 0.4 mA each, while the bases have 0.7 V. The 0.4 mA sourcing transistor's reverse saturation current must also be accounted for, as shown.

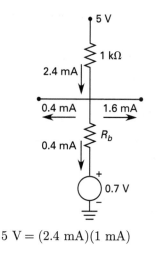

$$5 \text{ V} = (2.4 \text{ mA})(1 \text{ mA})$$
$$+ (0.4 \text{ mA})R_b + 0.7 \text{ V}$$

$$R_b = \frac{5 \text{ V} - 0.7 \text{ V} - 2.4 \text{ V}}{(0.4 \text{ mA})\left(0.001 \ \dfrac{\text{A}}{\text{mA}}\right)}$$

$$= \boxed{4750 \ \Omega \quad (4.75 \text{ k}\Omega)}$$

27.
$$\tau_{min} = \tau_d + \tau_r + \tau_s + \tau_f = 115 \text{ ns}$$
$$= 5 \text{ ns} + 50 \text{ ns} + 30 \text{ ns} + 30 \text{ ns}$$

$$f_{max} = \frac{1}{\tau_{min}} = \frac{10^9 \ \dfrac{\text{s}}{\text{ns}}}{115 \text{ ns}}$$

$$= \boxed{8.7 \times 10^6 \text{ Hz} \quad (8.7 \text{ MHz})}$$

28. The desired conversion is

The required circuit is an integrator, shown in the following illustration.

Use Kirchoff's current law (KCL) at the inverting terminal to get

$$\frac{0 \text{ V} - v_{in}}{R} + C \frac{d(0 \text{ V} - v_{out})}{dt} = 0$$

Rearranging and integrating gives

$$v_{out} = -\frac{1}{RC}\int v_{in}\,dt + \kappa$$

At half the period, 5 ms, the necessary initial conditions are the level of the voltage and the time frame.

$$v_{out} = -\frac{1}{RC}\int_{0\,s}^{0.005\,s} 5\,dt + 5\text{ V} = -5\text{ V}$$

$$= -\frac{5\text{ V}}{RC}t\bigg|_{0\,s}^{0.005\,s} + 5\text{ V} = -5\text{ V}$$

$$= -\frac{5\text{ V}}{RC}(0.005\text{ s}) = -10\text{ V}$$

Solve for the combination RC.

$$-\frac{5\text{ V}}{RC}(0.005\text{ s}) = -10\text{ V}$$

$$RC = -\frac{(5\text{ V})(0.005\text{ s})}{-10\text{ V}}$$

$$= \boxed{2.5 \times 10^{-3}\text{ s}}$$

Therefore, choosing an RC combination value equal to the calculated value provides the transient necessary to effect the desired conversion.

29. With both diodes open,

$$\frac{v_{out}}{v_{in}} = -1$$

With the upper diode conducting,

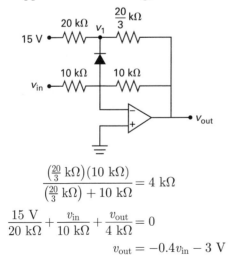

$$\frac{\left(\frac{20}{3}\text{ k}\Omega\right)(10\text{ k}\Omega)}{\left(\frac{20}{3}\text{ k}\Omega\right) + 10\text{ k}\Omega} = 4\text{ k}\Omega$$

$$\frac{15\text{ V}}{20\text{ k}\Omega} + \frac{v_{in}}{10\text{ k}\Omega} + \frac{v_{out}}{4\text{ k}\Omega} = 0$$

$$v_{out} = -0.4v_{in} - 3\text{ V}$$

For the lower diode conducting,

$$v_{out} = -0.4v_{in} + 3\text{ V}$$

For both diodes open,

$$v_1 = (15\text{ V})\left(\frac{20\text{ k}\Omega}{\frac{20}{3}\text{ k}\Omega + 20\text{ k}\Omega}\right)$$

$$+ v_{out}\left(\frac{20\text{ k}\Omega}{\frac{20}{3}\text{ k}\Omega + 20\text{ k}\Omega}\right)$$

$$= \frac{(15\text{ V})\left(\frac{20}{3}\text{ k}\Omega\right) + v_{out}(20\text{ k}\Omega)}{\frac{20}{3}\text{ k}\Omega + 20\text{ k}\Omega}$$

$$= \frac{15\text{ V} + 3v_{out}}{4}$$

$$= \frac{15\text{ V} - 3v_{in}}{4}$$

Because the negative terminal is at virtual ground, for the upper diode to be off, $v_1 > 0$ V, so

$$15\text{ V} - 3v_{in} > 0\text{ V}$$

$$v_{in} < 5\text{ V}$$

For the lower diode,

For both diodes open,

$$v_2 = (-15\text{ V})\left(\frac{\frac{20}{3}\text{ k}\Omega}{\frac{20}{3}\text{ k}\Omega + 20\text{ k}\Omega}\right)$$

$$- v_{in}\left(\frac{\frac{20}{3}\text{ k}\Omega}{\frac{20}{3}\text{ k}\Omega + 20\text{ k}\Omega}\right)$$

$$= \frac{(-15\text{ V})\left(\frac{20}{3}\text{ k}\Omega\right) - v_{in}(20\text{ k}\Omega)}{\frac{20}{3}\text{ k}\Omega + 20\text{ k}\Omega}$$

$$= \frac{-15\text{ V} - 3v_{in}}{4} < 0\text{ V}$$

$$v_{in} > -5\text{ V}$$

For -5 V $< v_{\text{out}} < 5$ V, $v_{\text{out}} = -v_{\text{in}}$

$$v_{\text{in}} < -5 \text{ V}, \quad v_{\text{out}} = -0.4 v_{\text{in}} + 3 \text{ V}$$

$$v_{\text{in}} > 5 \text{ V}, \quad v_{\text{out}} = -0.4 v_{\text{in}} - 3 \text{ V}$$

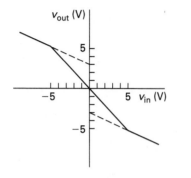

30. With at least one input low, the collector is high and no current flows in the collector circuit.

One or more low inputs of 0.2 V, when combined with a diode drop of 0.7 V, results in a current through the left resistor of

$$I = \frac{\Delta V}{R}$$

$$I = \frac{4 \text{ V} - 0.7 \text{ V} - 0.2 \text{ V}}{(2 \text{ k}\Omega)\left(1000 \; \frac{\Omega}{\text{k}\Omega}\right)}$$

$$= 1.55 \times 10^{-3} \text{ A}$$

The power for high output is

$$P = IV$$

$$= (1.55 \times 10^{-3} \text{ A})(4 \text{ V})$$

$$= 6.2 \times 10^{-3} \text{ W}$$

With both inputs high,

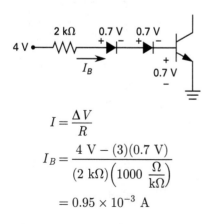

$$I = \frac{\Delta V}{R}$$

$$I_B = \frac{4 \text{ V} - (3)(0.7 \text{ V})}{(2 \text{ k}\Omega)\left(1000 \; \frac{\Omega}{\text{k}\Omega}\right)}$$

$$= 0.95 \times 10^{-3} \text{ A}$$

This saturates the transistor, so

$$I = \frac{\Delta V}{R}$$

$$I_C = \frac{4 \text{ V} - 0.2 \text{ V}}{(2 \text{ k}\Omega)\left(1000 \; \frac{\Omega}{\text{k}\Omega}\right)}$$

$$= 1.9 \times 10^{-3} \text{ A}$$

The power supply provides

$$P_{\text{supply}} = IV = (I_C + I_B)V$$

$$= (1.9 \times 10^{-3} \text{ A} + 0.95 \times 10^{-3} \text{ A})(4 \text{ V})$$

$$= 11.4 \times 10^{-3} \text{ W}$$

However, the transistor must absorb power from the loads, which must be calculated from a current of five times the input current for the high output. This excess collector current is

$$I_c = 5I = (5)(1.55 \times 10^{-3} \text{ A}) = 7.75 \times 10^{-3} \text{ A}$$

Assuming the transistor remains saturated, the excess transistor power can be as high as

$$P_{\text{excess}} = IV$$

$$= (7.75 \times 10^{-3} \text{ A})(0.2 \text{ V})$$

$$= 1.55 \times 10^{-3} \text{ W}$$

The maximum total power is

$$P_{\text{total}} = P_{\text{supply}} + P_{\text{excess}}$$

$$= 1.55 \times 10^{-3} \text{ W} + 11.4 \times 10^{-3} \text{ W}$$

$$= \boxed{12.95 \times 10^{-3} \text{ W} \quad (12.95 \text{ mW})}$$

47 Digital Logic

PRACTICE PROBLEMS

1. What is the state of a TTL cell at 4.0 V?

   (A) 0

   (B) indeterminate

   (C) 1

   (D) saturated

2. A computer accesses a register in 4-bit packets, sometimes called *nibbles*. What is the total number of values that can be represented in a nibble?

   (A) 4

   (B) 8

   (C) 15

   (D) 16

3. A computer data sheet indicates a 64K byte RAM. How many bytes of memory are available?

   (A) 8 bytes

   (B) 1024 bytes

   (C) 64,000 bytes

   (D) 65,536 bytes

4. How many bits of information are contained within a 64K byte RAM?

   (A) 64 bits

   (B) 8192 bits

   (C) 512,000 bits

   (D) 524,288 bits

5. If $x_1 = 0$, $x_2 = 1$, and $x_3 = 0$, what is the value of the following expression?

$$f(x_1, x_2, x_3) = \overline{x}_1 x_2 x_3 + x_1 \overline{x}_2 x_3 \cdot x_1 x_2 \overline{x}_3$$

6. A four-variable function has a minterm in row five. That minterm is given by which of the following expressions?

   (A) $\overline{A}B\overline{C}$

   (B) $A\overline{B}C$

   (C) $\overline{A}B\overline{C}D$

   (D) $A\overline{B}C\overline{D}$

The following truth table is applicable to Prob. 7 and Prob. 8.

A	B	C	$F(A,B,C)$
0	0	0	0
0	0	1	0
0	1	0	0
0	1	1	1
1	0	0	0
1	0	1	0
1	1	0	0
1	1	1	1

7. What is the canonical sum-of-products (SOP) expression for the function in the truth table shown?

   (A) $\overline{A}\,\overline{B}\,\overline{C} + \overline{A}\,\overline{B}C + \overline{A}B\overline{C}$

   (B) $A\overline{B}\,\overline{C} + A\overline{B}\,\overline{C} + AB\overline{C}$

   (C) (A) and (B) combined

   (D) $\overline{A}BC + ABC$

8. What maxterms are necessary to create the canonical product-of-sums (POS) expression for the function in the truth table shown?

   (A) $M_0 M_1 M_2$

   (B) $M_4 M_5 M_6$

   (C) (A) and (B) combined

   (D) $M_3 M_7$

The following truth table is applicable to Prob. 9 and Prob. 10.

A	B	$F(A,B)$
-5 V	-5 V	-5 V
-5 V	0	0
0	-5 V	0
0	0	-5 V

9. If positive logic is applied to the truth table, what logic operation is obtained?

(A) coincidence

(B) NOR

(C) XOR

(D) XNOR

10. If negative logic is applied to the truth table, what logic operation is obtained?

(A) coincidence

(B) NOR

(C) OR

(D) XOR

SOLUTIONS

1. The state of a cell is its binary value. For TTL, any voltage between 2.0 V and 5.0 V is state 1.

The answer is (C).

2. The total number of values in any n-tuple logic expression is given by Eq. 47.1.

$$N = 2^n = 2^4 = \boxed{16}$$

The answer is (D).

3. The K indicates $2^{10} = 1024$. So,

$$64\text{K} = (64)(2)^{10} = \boxed{65{,}536 \text{ bytes}}$$

The answer is (D).

4. One byte is 8 bits. Therefore,

$$(65{,}536 \text{ bytes})\left(8 \, \frac{\text{bits}}{\text{byte}}\right) = \boxed{524{,}288 \text{ bits}}$$

The answer is (D).

5. The function is

$$f(x_1, x_2, x_3) = \overline{x}_1 x_2 x_3 + x_1 \overline{x}_2 x_3 \cdot x_1 x_2 \overline{x}_3$$

Substituting the given values and using the order of precedence rules gives

$$\begin{aligned}
f(0,1,0) &= (\overline{0} \cdot 1 \cdot 0) + (0 \cdot \overline{1} \cdot 0) \cdot (0 \cdot 1 \cdot \overline{0}) \\
&= (1 \cdot 1 \cdot 0) + (0 \cdot 0 \cdot 0) \cdot (0 \cdot 1 \cdot 1) \\
&= 0 + 0 \cdot 0 \\
&= 0 + 0 \\
&= \boxed{0}
\end{aligned}$$

6. A function with four variables (that is, with $n=4$) with a value of one in row five has a minterm in row five. Changing row five from decimal to binary gives

$$5_{10} = 0101_2$$

The minterm is deduced from its binary representation as

$$\boxed{\overline{A} B \overline{C} D = m_5}$$

The answer is (C).

7. The minterms are in rows three and seven—that is, where the function equals one. The canonical sum-of-products form is

$$F(A, B, C) = m_3 + m_7 = \boxed{\overline{A}BC + ABC}$$

The answer is (D).

8. The maxterms are in the rows where the function has a value of zero.

$$F(A, B, C) = \boxed{\prod M(0, 1, 2, 4, 5, 6)}$$

The answer is (C).

9. If positive logic is applied, the high voltage has a value of one. Rewriting the truth table with -5 V as logic 0 and 0 V as logic 1 gives

A	B	$F(A,B)$
0	0	0
0	1	1
1	0	1
1	1	0

This is the exclusive OR (XOR) operation.

The answer is (C).

10. If negative logic is applied, the low voltage has a value of one. Rewriting the truth table with -5 V as logic 1 and 0 V as logic 0 gives

A	B	$F(A,B)$
1	1	1
1	0	0
0	1	0
0	0	1

Reversing the order to place the table in standard format gives

A	B	$F(A,B)$
0	0	1
0	1	0
1	0	0
1	1	1

This is the XNOR, or coincidence, logic.

The answer is (A).

Electronics

48 Lightning Protection and Grounding

PRACTICE PROBLEMS

1. What range of voltage does lightning span?

(A) 1 MV to 2 MV

(B) 2 MV to 5 MV

(C) 5 MV to 10 MV

(D) 10 MV to 1000 MV

2. What range of current does lightning span?

(A) 100 A to 1000 A

(B) 1000 A to 10,000 A

(C) 1000 A to 100,000 A

(D) 1000 A to 200,000 A

3. What is the name for an insulation level specified in terms of the crest value of the standard lightning impulse?

(A) BIL

(B) lightning impulse level

(C) TIL

(D) withstand voltage

4. A power system designed for what voltage is considered protected from direct lightning strikes by its own insulation?

(A) 15 kV

(B) 230 kV

(C) 330 kV

(D) none of the above

5. An arrester provides a protection level of 38 kV. If the minimum protective margin allowed by standards is 20%, is equipment with a 95 kV withstand-voltage adequately protected?

SOLUTIONS

1. Lightning voltage varies widely but is in the range of 10 MV to 1000 MV.

The answer is (D).

2. The current ranges from 1000 A to 200,000 A.

The answer is (D).

3. The voltage level that insulation is able to withstand, specified in terms of a standard lightning impulse, is called the basic lightning impulse level (BIL).

The answer is (A).

4. Systems of 230 kV or less cannot be insulated against the overvoltage caused by direct lightning strikes and must use a combination of shielding and grounding.

The answer is (C).

5. The protective margin is

$$\text{PM} = \frac{V_w - V_p}{V_p}$$

$$= \frac{95 \times 10^3 \text{ V} - 38 \times 10^3 \text{ V}}{38 \times 10^3 \text{ V}} \times 100\%$$

$$= 150\%$$

$$> 20\%$$

The equipment is adequately protected.

49 Illumination

PRACTICE PROBLEMS

1. What is the wavelength of a 60 Hz signal (using vacuum as the reference)?

(A) 0.016 m

(B) 60 m

(C) 5.0×10^6 m

(D) 3.0×10^8 m

2. What is the energy contained in a single quantum of visible light at a wavelength of 5.55×10^{-7} m?

(A) 3.7×10^{-40} J

(B) 3.6×10^{-19} J

(C) 5.6×10^{-7} J

(D) 170 J

3. Which type of light may be used to destroy bacteria?

(A) infrared

(B) laser

(C) visible

(D) ultraviolet

4. Which of the following statements about visible light is NOT true?

(A) The index of refraction is 1.5 in air.

(B) Visible light consists of many colors.

(C) The average frequency of visible light is approximately 5.4×10^{14} Hz.

(D) The wavelengths are approximately 380–780 nm.

5. Kirchhoff's law of thermal radiation specifies that the emissitivity, ϵ, equals the spectral absorption, α, in

(A) a blackbody

(B) a greybody

(C) a selective radiator

(D) none of the above

6. What is the maximum wavelength expected for a light source at 3000K?

(A) 9.7×10^{-7} m

(B) 3.3×10^{-4} m

(C) 2.9×10^{-3} m

(D) 3.0×10^{-3} m

7. What wavelength of light is emitted if an electron experiences a 2.2 V transition in a semiconductor?

(A) 220 nm

(B) 300 nm

(C) 555 nm

(D) 560 nm

8. The maximum lumens available per watt at a wavelength of 555 nm is _____ lm, though an ideal white source provides a maximum of _____ lm per watt.

(A) 555, 220

(B) 683, 220

(C) 683, 683

(D) 683, 432

9. The illuminance required at desktop level for a given task is 500 lx. Manufacturer data for the light source indicates an illuminance of 3500 lx at 1 m. The situation is as shown.

What is the illuminance level at the desk top?

(A) 80 lx

(B) 560 lx

(C) 1400 lx

(D) 14 000 lx

10. Consider the lighting configuration shown. The light sources all have the same intensity, $I = 500$ cd. What is the contribution of light source 2 at point A?

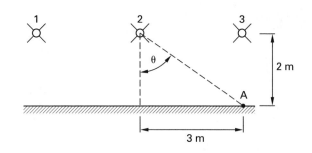

(A) 21 lx

(B) 42 lx

(C) 70 lx

(D) 130 lx

11. The required light level at a 36-inch high work plane is 50 fc. The luminaires hang 6 inches from the ceiling and have a coefficient of utilization (CU) of 0.8. The entire room is considered a work area. The room is 20 ft × 20 ft. What is most nearly the total lumens required from the luminaires to achieve the desired light level? Do not consider light loss factors.

(A) 50 lm

(B) 150 lm

(C) 6200 lm

(D) 25,000 lm

SOLUTIONS

1. The wavelength (with an index of refraction equal to one—that of a vacuum) is derived from the relation of the speed of light, c, to its wavelength, λ, and its frequency, ν.

$$c = \lambda \nu$$

$$\lambda = \frac{c}{\nu} = \frac{2.9979 \times 10^8 \ \frac{\text{m}}{\text{s}}}{60 \ \text{Hz}}$$

$$= \boxed{4.9665 \times 10^6 \ \text{m} \quad (5 \times 10^6 \ \text{m})}$$

The wavelength is 5000 km, or approximately 3100 mi. Because the index of refraction in air is nearly equal to the refraction index in a vacuum, this is also the wavelength in air. Extremely low frequency (ELF) waves (such as a 60 Hz signal) travel in the cavity between the Earth's surface and the ionosphere.

The answer is (C).

2. The human retina is most sensitive to visible light at 555 nm (5.55×10^{-7} m). Using Planck's constant, h, the energy of a single quantum of light at this wavelength is

$$E = h\nu = \frac{hc}{\lambda}$$

$$= \frac{(6.6256 \times 10^{-34} \ \text{J·s})\left(2.9979 \times 10^8 \ \frac{\text{m}}{\text{s}}\right)}{5.55 \times 10^{-7} \ \text{m}}$$

$$= \boxed{3.56 \times 10^{-19} \ \text{J} \quad (3.6 \times 10^{-19} \ \text{J})}$$

The answer is (B).

3. Ultraviolet light causes numerous biological effects, one of which is the destruction of various forms of bacteria.

The answer is (D).

4. The wavelengths of visible light are approximately 380–780 nm. Visible light is polychromatic, that is, it consists of many colors. The human retina is most sensitive to light at 555 nm. The index of refraction in air is approximately one. (The actual difference from one is so slight that it is unimportant in most calculations.)

The answer is (A).

5. A blackbody radiator has the property that

$$\epsilon = \alpha = 1$$

The answer is (A).

6. According to the Wien displacement law, the maximum wavelength depends only on the temperature, and is given by

$$\lambda_{\max} = \frac{b}{T}$$
$$= \frac{2.8978 \times 10^{-3} \text{ m·K}}{3000\text{K}}$$
$$= \boxed{9.7 \times 10^{-7} \text{ m}}$$

The constant b is determined by experiment.

The answer is (A).

7. The wavelength of emitted radiation in terms of the potential difference is

$$\lambda_{\text{nm}} = \frac{1239.76 \text{ nm·V}}{\Delta V}$$
$$= \frac{1239.76 \text{ nm·V}}{2.2 \text{ V}}$$
$$= \boxed{563 \text{ nm} \quad (560 \text{ nm})}$$

The answer is (D).

8. The theoretical maximum number of lumens available per watt of energy is 683 lm. Because of inefficiencies and unavoidable losses, an ideal white source can only provide 220 lm.

The answer is (B).

9. The inverse square law gives the illuminance from a point source at a straight-line distance, r, from the source as

$$E_1 r_1^2 = E_2 r_2^2$$
$$E_2 = \frac{E_1 r_1^2}{r_2^2}$$
$$= \frac{(3500 \text{ lx})(1 \text{ m})^2}{(2.5 \text{ m})^2}$$
$$= \boxed{560 \text{ lx}}$$

The answer is (B).

10. By the cosine-cubed law, the illuminance E, for a light source of intensity I, at point A, is

$$E = \frac{I \cos^3 \theta}{h^2} \qquad \text{[I]}$$

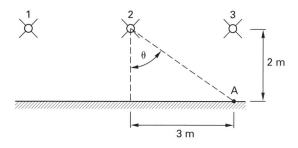

The variable h is the height. The only unknown is the angle θ.

$$\theta = \arctan \frac{\text{opposite}}{\text{adjacent}}$$
$$= \arctan \frac{3 \text{ m}}{2 \text{ m}}$$
$$= 56.3°$$

Substitute the value of θ into Eq. [I].

$$E = \frac{I \cos^3 \theta}{h^2}$$
$$= \frac{(500 \text{ cd})\cos^3 56.3°}{(2 \text{ m})^2}$$
$$= \boxed{21.35 \frac{\text{cd}}{\text{m}^2} \quad (21 \text{ lx})}$$

The answer is (A).

11. The light level is given by the following equation. The luminance is labeled initial since light loss factors (LLF) were not considered.

$$E_{\text{initial}} = \frac{L_{\text{total}}\text{CU}}{A_{\text{workplane}}}$$

Rearrange the equation to solve for the total lamp lumens, L_{total}. A footcandle (fc) is one lumen per square foot.

$$L_{\text{total}} = \frac{E_{\text{initial}}A_{\text{workplane}}}{\text{CU}}$$
$$= \frac{(50 \text{ fc})(20 \text{ ft})(20 \text{ ft})}{0.8}$$
$$= 25{,}000 \text{ lm}$$

The answer is (D).

Special Applications

50 Power System Management

PRACTICE PROBLEMS

1. A SCADA system configuration uses input from an instrument CT with a turns ratio of 100:5. The transducer supplying the SCADA system has a full-scale AC current value of 2.5 A. What is most nearly the maximum current in the primary side of the CT?

(A) 20 A

(B) 25 A

(C) 50 A

(D) 70 A

2. When generators are in parallel, their loads are shared. What electrical parameter is used to control the reactive load sharing between generators?

(A) current

(B) frequency

(C) speed

(D) voltage

3. Stability studies are used to determine the impact of disturbances on an electrical system. Which of the following assumptions are common to nearly all stability studies?

(A) Unbalanced faults are calculated using symmetrical components.

(B) Generated voltage is not affected by the speed of the prime mover.

(C) Harmonics are ignored.

(D) All of the above are common.

4. A main function of the SCADA system is to ensure economical operation of the electrical distribution system. This function is accomplished by evaluating the load, load forecast, fuel cost, and usage for each particular type of machine. What is the term for this function?

(A) energy management

(B) Modbus® adjustment

(C) power flow study

(D) scheduling

5. A pair of single-phase transformers is placed in parallel. (The transformers have the same voltage ratings.) The applicable data for the two transformers follows.

$$T_1: \quad S_1 = 2000 \text{ kVA}$$
$$Z_{T_1} = 3\%$$
$$T_2: \quad S_2 = 4000 \text{ kVA}$$
$$Z_{T_2} = 4.5\%$$

What is most nearly the maximum load allowed without overloading either of the transformers?

(A) 2000 kVA

(B) 3500 kVA

(C) 4600 kVA

(D) 6000 kVA

SOLUTIONS

1. SCADA refers to the Supervisory Control and Data Acquisition subsystem of the EMS. The relationship for an ideal transformer's primary and secondary currents (which will provide the maximum current) is given by

$$\frac{I_{\text{primary}}}{I_{\text{secondary}}} = \frac{n_{\text{secondary}}}{n_{\text{primary}}}$$

Rearrange the equation to solve for the primary current.

$$I_{\text{primary}} = \left(\frac{n_{\text{secondary}}}{n_{\text{primary}}}\right) I_{\text{secondary}}$$
$$= \left(\frac{100}{5}\right)(2.5 \text{ A})$$
$$= 50 \text{ A}$$

The answer is (C).

2. Adjusting the no-load frequency set point of a generator controls real load. Adjusting the no-load voltage set point of a generator controls the reactive load.

The answer is (D).

3. Unbalanced faults are calculated using the method of symmetrical components. Generated voltage is assumed to be constant in stability studies, so small changes in the speed of the prime mover are considered not to impact the voltage. DC offsets and harmonics are ignored. That is, only synchronous frequency voltages and currents are considered.

The answer is (D).

4. Energy management is a broad term applying to any aspect of power system management.

A Modbus is a serial communications protocol common to many SCADA systems.

A power flow study is used to determine the stability of a system.

Scheduling is the process by which the SCADA system evaluates the load, load forecast, fuel cost, usage, and numerous other factors to determine the most economical usage.

The answer is (D).

5. Use the per unit system to convert the impedance values of the first transformer to the second transformer. The voltage base is the same on both transformers.

$$Z_{\text{pu,new}} = Z_{\text{pu,old}} \left(\frac{V_{\text{base,old}}}{V_{\text{base,new}}}\right)^2 \left(\frac{S_{\text{base,new}}}{S_{\text{base,old}}}\right)$$
$$= j0.03(1)^2 \left(\frac{4000 \text{ kVA}}{2000 \text{ kVA}}\right)$$
$$= j0.06$$

The transformers are now on the same base. The voltage base, regardless of the exact value of that voltage, is the same for both transformers. Taking the ratio of the full power gives

$$\frac{S_{T_1}}{S_{T_2}} = \frac{\dfrac{V^2}{Z_{T_1}}}{\dfrac{V^2}{Z_{T_2}}} = \frac{Z_{T_2}}{Z_{T_1}}$$

The transformers will add load in inverse proportion to their impedances. Substitute the same base values and solve for the full power of T_2 when T_1 is fully loaded.

$$\frac{S_{T_1}}{S_{T_2}} = \frac{Z_{T_2}}{Z_{T_1}}$$
$$S_{T_2} = S_{T_1}\left(\frac{Z_{T_1}}{Z_{T_2}}\right) = (2000 \text{ kVA})\left(\frac{0.06}{0.045}\right)$$
$$= 2667 \text{ kVA}$$

The total load of the transformers is

$$S_{\text{total}} = S_{T_1} + S_{T_2}$$
$$= 2000 \text{ kVA} + 2667 \text{ kVA}$$
$$= 4667 \text{ kVA} \quad (4600 \text{ kVA})$$

The answer is (C).

51 Measurement and Instrumentation

PRACTICE PROBLEMS

1. What instrument term can be defined as a measure of the reproducibility of observations?

(A) accuracy

(B) error

(C) precision

(D) resolution

2. A digital voltmeter is designed for a maximum ambient temperature of 65°C. On the 10 kHz scale, what is the expected noise voltage in a 100 kΩ series resistor?

(A) 4.3 nV

(B) 0.14 μV

(C) 1.9 μV

(D) 4.3 μV

3. What is the average value of the voltage signal $v(t) = 5 \sin 377t$?

(A) 0.0 V

(B) 3.2 V

(C) 3.5 V

(D) 5.6 V

4. The voltage signal in Prob. 3 is rectified and measured by a d'Arsonval meter movement. What is the voltage reading?

(A) 0.0 V

(B) 3.2 V

(C) 3.5 V

(D) 5.6 V

5. The manufacturer's data sheet lists the sensitivity of a DC voltmeter as 1000 Ω/V. What current is required for full-scale meter deflection?

(A) 1.0 nA

(B) 1.0 μA

(C) 1.0 mA

(D) 1.0 A

6. A DC voltmeter with a coil resistance of 150 Ω is to be designed with a sensitivity of 10,000 Ω/V and a voltage range of 0 V to 100 V. What type and what magnitude of resistance are required?

(A) parallel, 10 kΩ

(B) parallel, 1 MΩ

(C) series, 10 kΩ

(D) series, 1 MΩ

7. A d'Arsonval movement with a coil resistance of 150 Ω is to be designed for use as an ammeter. What is the full-scale deflection current?

(A) 0.3 mA

(B) 0.3 A

(C) 3.0 A

(D) none of the above

8. A DC voltmeter with 20 kΩ of internal resistance measures the voltage across a resistance as 10.0 V. The current through the parallel combination of the resistor and voltmeter is measured as 1.00 mA by an ammeter with 100 Ω of internal resistance. The resistor is fed from a source with a Thevenin resistance of 600 Ω. Determine the (a) resistance of the resistor and (b) current that will flow in the resistor if the meters are removed from the circuit.

9. A voltmeter is designed to measure voltages in the full-scale ranges of 3, 10, 30, and 100 V_{DC}. The meter movement has an internal resistance of 50 Ω and a full-scale current of 1 mA. Using a four-contact multiposition switch, design the voltmeter and specify the four resistor values used.

10. A digital meter uses the number system base b and has n digits. The noise is one-half the place value of the least significant digit. (a) Determine the signal-to-noise ratio for the case where the reading is half the full-scale value. (b) Obtain the signal-to-noise ratios for the following cases.

n	b	S/N (dB)
3	10	$20 \log (\text{S/N})$
4	8	
8	2	
3	16	

SOLUTIONS

1. Precision is the range of values of a set of measurements. Therefore, it is a measure of the reproducibility of observations.

The answer is (C).

2. The thermal noise voltage is

$$V_{\text{noise}} = \sqrt{4\kappa T R(\text{BW})}$$

$$= \sqrt{\begin{array}{c}(4)\left(1.3805 \times 10^{-23} \, \frac{\text{J}}{\text{K}}\right)(338 \text{ K}) \\ \times (100 \times 10^3 \, \Omega)(10 \times 10^3 \text{ Hz})\end{array}}$$

$$= \boxed{4.3 \times 10^{-6} \text{ V} \quad (4.3 \ \mu\text{V})}$$

The answer is (D).

3. The average value of a sinusoid is zero.

The answer is (A).

4. A d'Arsonval meter responds to average values. The average value of a rectified sinusoid is

$$V_{\text{ave}} = \left(\frac{2}{\pi}\right)V_p$$

$$= \left(\frac{2}{\pi}\right)(5 \text{ V})$$

$$= \boxed{3.18 \text{ V} \quad (3.2 \text{ V})}$$

The answer is (B).

5. The sensitivity is defined as

$$\text{sensitivity} = \frac{1}{I_{\text{fs}}}$$

$$I_{\text{fs}} = \frac{1}{\text{sensitivity}} = \frac{1}{1000 \, \frac{\Omega}{\text{V}}}$$

$$= \boxed{10^{-3} \text{ A} \quad (1.0 \text{ mA})}$$

The answer is (C).

6. The voltmeter requires an external series resistance given by

$$\frac{1}{I_{\text{fs}}} = \frac{R_{\text{ext}} + R_{\text{coil}}}{V_{\text{fs}}}$$

$$R_{\text{ext}} = V_{\text{fs}}\left(\frac{1}{I_{\text{fs}}}\right) - R_{\text{coil}}$$

$$= (100 \text{ V})\left(10{,}000 \, \frac{\Omega}{\text{V}}\right) - 150 \ \Omega$$

$$= \boxed{0.999 \times 10^6 \ \Omega \quad (1 \text{ M}\Omega)}$$

The answer is (D).

7. For ammeters, the standard design voltage is 50 mV. The full-scale deflection current is

$$I_{\text{fs}} = \frac{V}{R_{\text{coil}}} = \frac{50 \times 10^{-3} \text{ V}}{150 \ \Omega}$$

$$= \boxed{3.33 \times 10^{-4} \text{ A} \quad (0.3 \text{ mA})}$$

The answer is (A).

8.

(a) Assuming the meters are perfectly accurate,

$$V_{\text{Th}} = 10 \text{ V} + (0.001 \text{ A})(700 \ \Omega) = 10.7 \text{ V}$$

With 1 mA flowing into 10 V, the equivalent load resistance is $(10 \text{ V}/1 \text{ mA}) = 10 \text{ k}\Omega$, so

$$10 \text{ k}\Omega = \frac{(20 \text{ k}\Omega)R}{20 \text{ k}\Omega + R}$$

$$(20 \text{ k}\Omega + R)(10 \text{ k}\Omega) = (20 \text{ k}\Omega)R$$

$$(200 \text{ M}\Omega)\left(1000 \, \frac{\text{k}\Omega}{\text{M}\Omega}\right) + (10 \text{ k}\Omega)R = (20 \text{ k}\Omega)R$$

$$200 \times 10^3 \text{ k}\Omega = (10 \text{ k}\Omega)R$$

$$R = \boxed{20 \text{ k}\Omega}$$

(b) Removing the meters, the current flowing is

$$I = \frac{10.7 \text{ V}}{20{,}600 \ \Omega} = \boxed{0.52 \times 10^{-3} \text{ A} \quad (0.52 \text{ mA})}$$

9. The meter configuration is shown.

With the 1 mA full-scale current,

$$3 \text{ V}: \quad 50 + R_1 = \frac{3 \text{ V}}{1 \times 10^{-3} \text{ A}} = 3000 \text{ }\Omega$$

$$R_1 = \boxed{2950 \text{ }\Omega}$$

$$10 \text{ V}: \quad R_2 = \frac{10 \text{ V} - 3 \text{ V}}{1 \times 10^{-3} \text{ A}}$$

$$= \frac{7 \text{ V}}{1 \times 10^{-3} \text{ A}}$$

$$= \boxed{7000 \text{ }\Omega}$$

$$30 \text{ V}: \quad R_3 = \frac{30 \text{ V} - 10 \text{ V}}{1 \times 10^{-3} \text{ A}}$$

$$= \frac{20 \text{ V}}{1 \times 10^{-3} \text{ A}}$$

$$= \boxed{20{,}000 \text{ }\Omega}$$

$$100 \text{ V}: \quad R_4 = \frac{100 \text{ V} - 30 \text{ V}}{1 \times 10^{-3} \text{ A}}$$

$$= \frac{70 \text{ V}}{1 \times 10^{-3} \text{ A}}$$

$$= \boxed{70{,}000 \text{ }\Omega}$$

10. (a) Full-scale is 999_{10} for the first case. The noise is 0.5. At half-scale the signal is 500_{10}, so the signal-to-noise ratio is $500/0.5 = 10^3$ or

$$\text{S/N}|_{\text{dB}} = 20 \log 10^3 = \boxed{60 \text{ dB}}$$

(b) For the octal, full-scale in base 10 is

$$(7)(8)^3 + (7)(8)^2 + (7)(8)^1 + 7 = 4095$$

Half-scale is 2048.

$$\text{S/N} = 20 \log \frac{2048}{0.5} = \boxed{72.2 \text{ dB}}$$

For the binary, full-scale in base 10 is

$$(2)^7 + (2)^6 + (2)^5 + (2)^4 + (2)^3 + (2)^2 + (2)^1 + 1 = 255$$

Half-scale is 128.

$$\text{S/N} = 20 \log \frac{128}{0.5} = \boxed{48.2 \text{ dB}}$$

For the hexadecimal, full-scale is

$$(15)(16)^2 + (15)(16)^1 + 15 = 4095$$

For half-scale, the S/N ratio is the same as the octal half-scale.

$$\text{S/N} = 20 \log \frac{2048}{0.5} = \boxed{72.2 \text{ dB}}$$

52 Electrical Materials

PRACTICE PROBLEMS

1. Which of the following types of material has high thermal and electrical conductivity?

(A) ceramic

(B) metal

(C) polymer

(D) semiconductor

2. Which of the following elements is a liquid at standard atmospheric temperature and pressure (STP)?

(A) Ge

(B) H

(C) Hg

(D) In

3. Which of the following materials has the highest conductivity at room temperature?

(A) aluminum

(B) copper

(C) gold

(D) silver

4. Which of the following terms does NOT describe a ceramic material?

(A) crystalline

(B) high electrical resistance

(C) high electrical conductivity

(D) insulator

5. Which type of bonding is responsible for the attraction of water molecules to one another?

(A) covalent

(B) ionic

(C) metallic

(D) van der Waals

SOLUTIONS

1. Several divisions of materials can occur during classification. One such division is into metals, polymers (that is, plastics), and ceramics. Metal has the highest thermal and electrical conductivity of the material types.

The answer is (B).

2. Of the elements, only bromine (Br) and mercury (Hg) are liquid under standard atmospheric conditions.

The answer is (C).

3. The approximate values for electrical conductivity at 20°C for the material options listed is shown in the table.

metal	S/m
aluminium	38×10^6
copper	58×10^6
gold	41×10^6
silver	63×10^6

The answer is (D).

4. Ceramic materials can be either crystalline or amorphous, exhibit poor toughness (that is, they tend to be brittle), and possess low electrical conductivity—hence their use as insulators.

The answer is (C).

5. Ionic, covalent, and metallic bonds are the three primary bonding methods. Bonds between molecules of the same substance are called van der Waals forces. There are three types of van der Waals forces: dipole-to-dipole (between polar molecules), dispersion (between non-polar molecules), and hydrogen bond (also a dipole-to-dipole bond, but stronger).

Therefore, the van der Waals bonding is responsible for the attraction of water molecules.

The answer is (D).

53 Biomedical Electrical Engineering

PRACTICE PROBLEMS

The following information is applicable to Prob. 1 and Prob. 2.

A medical incubator is designed to maintain temperature, humidity, and CO_2 at the proper levels for the sampled items. The humidifier uses the low-water-level alarm circuitry shown.

1. What term describes the configuration of the two transistors?

(A) common collector

(B) common base

(C) common source

(D) Darlington pair

2. Which of the following transistor parameters represents the gain used to drive the control relay?

(A) h_{fe}

(B) h_{ie}

(C) h_{oe}

(D) h_{re}

3. A Class II biological safety cabinet (hood) uses a 115 V single-phase AC motor to maintain the vacuum level within the hood necessary to prevent the spread of biological contaminants. With clean filters, the motor draws 10 A at its rated power factor of 0.85 lagging. What is the power consumption of the motor?

(A) 980 W

(B) 1150 W

(C) 1700 W

(D) 1990 W

4. Leakage current standards for biomedical devices are generally in the microamp (μA) range. Which of the following items are used to reduce leakage currents for 60 Hz systems?

(A) DC-DC inverters

(B) rechargeable batteries

(C) motor-generators

(D) all of the above

5. Radiofrequency units used for patient care may result in _____ human tissue.

(A) reorientation of

(B) heating of

(C) increased circulation within

(D) reorganization of cellular structure of

SOLUTIONS

1. The two transistors are in a common emitter configuration (analogous to a common source configuration for field-effect transistors (FETs)). This specific configuration is more commonly referred to as a Darlington pair.

The answer is (D).

2. The Darlington pair has a high input resistance and a high current gain (h_{fe}). It is this forward current gain that enables operation of the control relay.

The answer is (A).

3. The power in a single-phase motor is

$$P = IV(\text{pf})$$
$$= (10 \text{ A})(115 \text{ V})(0.85)$$
$$= \boxed{977.5 \text{ W} \quad (980 \text{ W})}$$

The answer is (A).

4. Leakage currents are minimized by using nonoscillatory power sources (such as DC-DC inverters and rechargeable batteries) or electrical items with nonelectromagnetic coupling (such as motor-generators).

The answer is (D).

5. Radiofrequency fields of sufficient intensity can damage human cells by deposition of energy through heating.

The answer is (B).

Codes and Standards

54 National Electrical Code

PRACTICE PROBLEMS

1. An unknown wire gage must be determined during an electrical inspection. The wire diameter is measured as 0.2591 cm. What is the AWG standard designation?

(A) AWG 4/0

(B) AWG 4

(C) AWG 8

(D) AWG 10

2. A given circuit is meant to carry a continuous lighting load of 16 A. In addition, four loads designed for permanent display stands are fastened in place and require 2 A each when operating. What is the rating of the overcurrent protective device (OCPD) on the branch circuit?

(A) 20 A

(B) 23 A

(C) 28 A

(D) 30 A

3. If standard three-conductor cable is used for the branch circuit in Prob. 2, what is the minimum conductor size?

(A) AWG 8

(B) AWG 10

(C) AWG 12

(D) AWG 14

4. A dwelling two-wire resistive heating circuit is designed for 240 V and 30 A. AWG 8 copper wire is used with a resistance of 0.809 Ω/1000 ft. What is the voltage drop for a one-way circuit length of 75 ft?

(A) 1.0%

(B) 1.5%

(C) 3.6%

(D) 5.0%

5. A three-wire feeder, AWG 1/0, carries 100 A of continuous load from a 480 V, 3ϕ (three-phase), three-wire source. The THHN copper conductors operate at 75°C and are enclosed within aluminum conduit. The expected power factor is 0.8. The total circuit length is 150 m. What is the voltage drop?

(A) 4.0 V

(B) 7.0 V

(C) 11 V

(D) 14 V

6. A 120 V dwelling branch circuit supplies four outlets, one of which has four receptacles. What is the total volt-ampere load for the circuit shown?

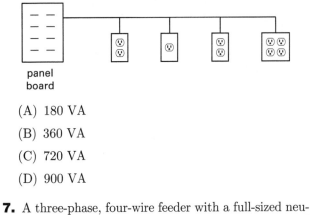

panel
board

(A) 180 VA

(B) 360 VA

(C) 720 VA

(D) 900 VA

7. A three-phase, four-wire feeder with a full-sized neutral in a separate raceway carries 14 A continuous and 40 A noncontinuous loads. The feeder cable has a rating of 75°C. What is the minimum copper conductor size?

(A) AWG 3

(B) AWG 4

(C) AWG 6

(D) AWG 8

Codes and Standards

8. All of the following are properties of fuses rather than circuit breakers, with one exception. Which item is the property of circuit breakers?

   (A) adjustable time-current characteristics

   (B) low initial costs

   (C) longer downtime after operation

   (D) simplicity

9. Installed equipment within a dwelling is attached to an overcurrent device (breaker) rated at 15 A. What is the required size of the equipment's grounding copper conductor?

   (A) AWG 10

   (B) AWG 12

   (C) AWG 14

   (D) AWG 18

10. A three-phase, four-wire feeder with termination provisions designed to carry less than 100 A uses THWN AWG 8 copper wire in an ambient operating environment of 50°C. From NEC Table 310.15(b)(2)(a), the derating factor for four conductors in a single raceway is 0.8. What is the allowable ampacity?

   (A) 19 A

   (B) 24 A

   (C) 30 A

   (D) 40 A

Codes and Standards

SOLUTIONS

1. The wire diameter is

$$d = (0.2591 \text{ cm})\left(\frac{1 \text{ in}}{2.54 \text{ cm}}\right) = 0.1020 \text{ in}$$

Finding this value in a table of wire gages, such as the one in App. 54.A, indicates a 10 gage wire. That is, AWG 10.

The answer is (D).

2. From Article 210.20(A) of the *National Electrical Code* (NEC), the overcurrent protective device (OCPD) must be rated at 100% of the noncontinuous load plus 125% of the continuous load.

$$\begin{aligned} OCPD &= (1.00)(2 \text{ A} + 2 \text{ A} + 2 \text{ A} + 2 \text{ A}) + (1.25)(16 \text{ A}) \\ &= 28 \text{ A minimum} \end{aligned}$$

A standard fixed-trip circuit breaker or a fuse rated at 30 A can be used (see Sec. 240.6).

The answer is (D).

3. Conductor size is based on 100% of the noncontinuous load plus 125% of the continuous load [Sec. 210.19 (A)(1)]. From Prob. 2, the total load is 28 A. The ampacity cannot be less than the rating of the overcurrent device, 30 A in this case [Sec. 310.15(B)FPN(2)]. The ampacity of the conductor for a circuit of less than 100 A should be based on the 60°C column of NEC Table 310.16 [Sec. 110.14(C)(1)] (see App. 54.B).

Using App. 54.B, an AWG 10 conductor is chosen. (Checking Article 240.3 of the *Code*, as is called for by the asterisk in App. 54.B, indicates that 10 gage wire cannot have overcurrent protection greater than 30 A. All requirements are met.) Any derating of the conductor ampacity requires the use of smaller gage (larger diameter) conductor. See the conductor factors at the bottom of App. 54.B.

The answer is (B).

4. A two-wire circuit is either single phase or DC. The heating is resistive, so pf = 1 if the circuit is AC, which is the most likely scenario. Regardless, the voltage drop is given by Ohm's law as

$$V = IR$$
$$= IR_l l$$

The voltage drop occurs on both wires, so a factor of two must be used.

$$V = IR_l 2l$$

Substituting gives

$$V = 2IR_l l$$

$$= (2)(30 \text{ A})\left(\frac{0.809 \ \Omega}{1000 \text{ ft}}\right)(75 \text{ ft})$$

$$= 3.64 \text{ V}$$

As a percentage of the 240 V nominal voltage, the voltage drop is

$$\left(\frac{3.64 \text{ V}}{240 \text{ V}}\right) \times 100\% = \boxed{1.516\% \quad (1.5\%)}$$

The answer is (B).

The conductor ampacities can be found in App. 54.B (NEC Table 310.16). Conductor properties, such as resistance (Ω) per 1000 ft, are found in App. 54.C (NEC Chap. 9, Table 8).

5. Although the continuous load is 100 A, and the voltage drop is calculated using that value, the conductors and the overcurrent protector must be able to carry 125 A, that is, 125% of the continuous load [Sec. 215-2(a) and Sec. 215-3].

The AC voltage drop is

$$V_{\text{drop}} = IZ$$

The effective impedance, corrected for the power factor, is calculated using App. 54.D (NEC Chap. 9, Table 9). For a 1/0 gage feeder, the resistance to neutral per 1000 ft in aluminum conduit is 0.13 Ω.

The inductive reactance, X_L, to neutral per 1000 ft is 0.044 Ω. The columns titled "alternating-current resistance for uncoated copper wires" and "X_L (reactance) for all wires" are used. The uncoated copper wire column is used, although the THHN conductors are insulated. Also, the capacitive reactance is ignored. Therefore, the voltage drop calculated will be an approximate value.

Because the power factor is not 0.85, the effective impedance cannot be taken directly from the table. The corrected impedance, Z_c, is

$$Z_c = R_l(\text{pf}) + X_{L,l} \sin(\arccos \text{pf})$$

$$= \left(\frac{0.13 \ \Omega}{1000 \text{ ft}}\right)(0.8)$$

$$\quad + \left(\frac{0.044 \ \Omega}{1000 \text{ ft}}\right) \sin 36.86°$$

$$= 0.1304 \ \Omega/1000 \text{ ft}$$

Because this is an "ohms-to-neutral" value, the line-to-neutral, or phase, voltage drop is

$$V_{\text{drop (line-to-neutral)}} = V_\phi = IZ_c$$

$$= (100 \text{ A})\left(\frac{0.1304 \ \Omega}{1000 \text{ ft}}\right)$$

$$= 13.04 \text{ V}/1000 \text{ ft}$$

Adjusting for the total circuit length gives

$$V_\phi = \left(\frac{13.04 \text{ V}}{1000 \text{ ft}}\right)(150 \text{ m})\left(\frac{1 \text{ ft}}{0.3048 \text{ m}}\right)$$

$$= 6.417 \text{ V}$$

Voltage drops are given in terms of line quantities for a three-phase, three-wire system.

$$V_{\text{line}} = V_\phi \sqrt{3}$$

$$= (6.417 \text{ V})\sqrt{3}$$

$$= \boxed{11.1 \text{ V} \quad (11 \text{ V})}$$

This is approximately 2.5% of the 480 V source and is within NEC recommendations.

The answer is (C).

6. The situation is as shown.

On any outlet with less than four receptacles, the VA load is 180 VA [Sec. 220.14(I)]. For four or more receptacles, the load is

$$\text{VA} = (\text{numbers of receptacles})\left(90 \ \frac{\text{VA}}{\text{receptacle}}\right)$$

$$= (4)(90 \text{ VA})$$

$$= 360 \text{ VA}$$

$$\text{VA}_{\text{total}} = 180 \text{ VA} + 180 \text{ VA} + 180 \text{ VA} + 360 \text{ VA}$$

$$= \boxed{900 \text{ VA}}$$

The answer is (D).

7. Feeder conductor size, before derating, is based on 100% of the noncontinuous load and 125% of the continuous load [Art. 215.2(A)(1)].

$$A_{\text{load}} = (1.00)(40 \text{ A}) + (1.25)(14 \text{ A})$$
$$= 40 \text{ A} + 17.5 \text{ A}$$
$$= 57.5 \text{ A} \quad (58 \text{ A})$$

The total of 58 A is used because an ampacity of 0.5 A or greater is rounded up [Sec. 220.5(B)]. Using App. 54.B (NEC Table 310.16), select AWG 6 from the 75°C column. Always consider additional requirements or limits on the ratings. In this case, for circuits less than 100 A, the 60°C column is required [Sec. 110.14(C)(1)]. (Because of this article, the column primarily used in dwelling calculations is the one titled "60°C (140°F)".) Therefore, the AWG 6 conductor cannot be used as it is rated for 55 A at 60°C. Instead, AWG 4 with a 70 A rating is used.

The answer is (B).

8. The electrical characteristics of a fuse are set by the material used, and no external adjustments can occur. The time-current characteristics cannot change.

The answer is (A).

9. Equipment grounding conductor sizes are given in NEC Table 250.122. For a 15 A overcurrent device, the grounding conductor must be AWG 14 copper.

The answer is (C).

10. Because the circuit carries less than 100 A and the termination provisions are designed commensurate with this condition, the 60°C column of App. 54.B must be used [see Sec. 110.14(C)(1)]. 40 A is the limiting allowable ampacity. The operating environment is above 30°C for which the values are calculated. THWN wire is rated for 75°C and 50 A. Using the correction factor table in App. 54.B, 0.75 is found in the 46–50°C row. The allowable ampacity in a given environment is determined by multiplying the rated ampacity in the table by all the correction factors—in this case, one for temperature and one for more than three conductors in a raceway.

$$A_{\text{total}} = A_{\text{rated}} F_{\text{temp}} F_{\text{cond}}$$
$$= (50 \text{ A})(0.75)(0.8)$$
$$= \boxed{30 \text{ A}}$$

Because 30 A is less than the 40 A maximum limit from the 60°C column, 30 A is the correct value.

The answer is (C).

Codes and Standards

55 National Electrical Safety Code

PRACTICE PROBLEMS

1. Street and area lights under the control of utilities are covered under the rules of the

(A) ANSI

(B) IEEE

(C) NEC

(D) NESC

2. What is the approximate required minimum clearance distance, R, if the phase voltage of the live parts is 13,800 V?

(A) 1 m

(B) 4 m

(C) 10 m

(D) 30 m

3. The New England states are considered to be in the _____ weather zone, according to Rule 250, and so must meet the clearance requirements of zone _____ of Rule 230.

(A) medium, 1

(B) medium, 2

(C) heavy, 1

(D) heavy, 2

4. What approximate vertical clearance is required for an open supply conductor at 13.8 kV over a body of water (> 8 km^2 surface area) suitable for sail boating?

(A) 4 m

(B) 8 m

(C) 12 m

(D) 30 m

5. The wind load in newtons, L_{SI}, on an overhead line structure is calculated using the following formula.

$$L_{SI} = 0.613\,V_{m/s}^2\,k_z\,G_{RF}\,IC_f\,A_{m^2}$$

What does the k_z term represent?

(A) force coefficient

(B) gust response factor

(C) velocity pressure coefficient

(D) wind speed

SOLUTIONS

1. The NESC (National Electrical Safety Code) contains the governing requirements for lights under the exclusive control of utilities (or their authorized contractors).

The answer is (D).

2. According to NESC Rule 110 and Table 110-1, for a voltage of 13,800 V, dimension R must be at least 3.1 m (4 m) or 10.1 ft (10 ft).

The answer is (B).

3. The New England states are in the heavy weather zone. The heavy, medium, and light weather conditions of Rule 250 correspond to clearance zones 1, 2, and 3, respectively, of Rule 230.

The answer is (C).

4. According to Rule 232, Table 232-1, of the NESC (row 7.d, column 5), the necessary clearance is 12.3 m (12 m) based on the high water mark.

The answer is (C).

5. The formula shown, along with the definitions of each term, is given in NESC Rule 250.C. The term k_z is the velocity pressure exposure coefficient (see Rule 250.C.1, Table 250-2).

The answer is (C).

Codes and Standards

56 Engineering Economic Analysis

PRACTICE PROBLEMS

1. At 6% effective annual interest, how much will be accumulated if $1000 is invested for ten years?

2. At 6% effective annual interest, what is the present worth of $2000 that becomes available in four years?

3. At 6% effective annual interest, how much should be invested to accumulate $2000 in 20 years?

4. At 6% effective annual interest, what year-end annual amount deposited over seven years is equivalent to $500 invested now?

5. At 6% effective annual interest, what will be the accumulated amount at the end of ten years if $50 is invested at the end of each year for ten years?

6. At 6% effective annual interest, how much should be deposited at the start of each year for ten years (a total of 10 deposits) in order to empty the fund by drawing out $200 at the end of each year for ten years (a total of 10 withdrawals)?

7. At 6% effective annual interest, how much should be deposited at the start of each year for five years to accumulate $2000 on the date of the last deposit?

8. At 6% effective annual interest, how much will be accumulated in ten years if three payments of $100 are deposited every other year for four years, with the first payment occurring at $t = 0$?

9. $500 is compounded monthly at a 6% nominal annual interest rate. How much will have accumulated in five years?

10. What is the effective annual rate of return on an $80 investment that pays back $120 in seven years?

11. A new machine will cost $17,000 and will have a resale value of $14,000 after five years. Special tooling will cost $5000. The tooling will have a resale value of $2500 after five years. Maintenance will be $2000 per year. The effective annual interest rate is 6%. What will be the average annual cost of ownership during the next five years?

12. An old covered wooden bridge can be strengthened at a cost of $9000, or it can be replaced for $40,000. The present salvage value of the old bridge is $13,000. It is estimated that the reinforced bridge will last for 20 years, will have an annual cost of $500, and will have a salvage value of $10,000 at the end of 20 years. The estimated salvage value of the new bridge after 25 years is $15,000. Maintenance for the new bridge would cost $100 annually. The effective annual interest rate is 8%. Which is the best alternative?

13. A firm expects to receive $32,000 each year for 15 years from sales of a product. An initial investment of $150,000 will be required to manufacture the product. Expenses will run $7530 per year. Salvage value is zero, and straight-line depreciation is used. The income tax rate is 48%. What is the after-tax rate of return?

14. A public works project has initial costs of $1,000,000, benefits of $1,500,000, and disbenefits of $300,000. (a) What is the benefit/cost ratio? (b) What is the excess of benefits over costs?

15. A speculator in land pays $14,000 for property that he expects to hold for ten years. $1000 is spent in renovation, and a monthly rent of $75 is collected from the tenants. (Use the year-end convention.) Taxes are $150 per year, and maintenance costs are $250 per year. What must be the sale price in ten years to realize a 10% rate of return?

16. What is the effective annual interest rate for a payment plan of 30 equal payments of $89.30 per month when a lump sum payment of $2000 would have been an outright purchase?

17. A depreciable item is purchased for $500,000. The salvage value at the end of 25 years is estimated at $100,000. What is the depreciation in each of the first three years using the (a) straight line, (b) sum-of-the-years' digits, and (c) double-declining balance methods?

18. Equipment that is purchased for $12,000 now is expected to be sold after ten years for $2000. The estimated maintenance is $1000 for the first year, but it is expected to increase $200 each year thereafter. The effective annual interest rate is 10%. What are the (a) present worth and (b) annual cost?

19. A new grain combine with a 20-year life can remove seven pounds of rocks from its harvest per hour. Any rocks left in its output hopper will cause $25,000 damage in subsequent processes. Several investments are available to increase the rock-removal capacity, as listed in the table. The effective annual interest rate is 10%. What should be done?

rock removal rate	probability of exceeding rock removal rate	required investment to achieve removal rate
7	0.15	$0
8	0.10	$15,000
9	0.07	$20,000
10	0.03	$30,000

20. (*Time limit: one hour*) A mechanism that costs $10,000 has operating costs and salvage values as given. An effective annual interest rate of 20% is to be used.

year	operating cost	salvage value
1	$2000	$8000
2	$3000	$7000
3	$4000	$6000
4	$5000	$5000
5	$6000	$4000

(a) What is the economic life of the mechanism? (b) Assuming that the mechanism has been owned and operated for four years already, what is the cost of owning and operating the mechanism for one more year?

21. (*Time limit: one hour*) A salesperson intends to purchase a car for $50,000 for personal use, driving 15,000 miles per year. Insurance for personal use costs $2000 per year, and maintenance costs $1500 per year. The car gets 15 miles per gallon, and gasoline costs $1.50 per gallon. The resale value after five years will be $10,000. The salesperson's employer has asked that the car be used for business driving of 50,000 miles per year and has offered a reimbursement of $0.30 per mile. Using the car for business would increase the insurance cost to $3000 per year and maintenance to $2000 per year. The salvage value after five years would be reduced to $5000. If the employer purchased a car for the salesperson to use, the initial cost would be the same, but insurance, maintenance, and salvage would be $2500, $2000, and $8000, respectively. The salesperson's effective annual interest rate is 10%. (a) Is the reimbursement offer adequate? (b) With a reimbursement of $0.30 per mile, how many miles must the car be driven per year to justify the employer buying the car for the salesperson to use?

22. (*Time limit: one hour*) Alternatives A and B are being evaluated. The effective annual interest rate is 10%. What alternative is economically superior?

	alternative A	alternative B
first cost	$80,000	$35,000
life	20 years	10 years
salvage value	$7000	0
annual costs		
years 1–5	$1000	$3000
years 6–10	$1500	$4000
years 11–20	$2000	0
additional cost		
year 10	$5000	0

23. (*Time limit: one hour*) A car is needed for three years. Plans A and B for acquiring the car are being evaluated. An effective annual interest rate of 10% is to be used. Which plan is economically superior?

Plan A: lease the car for $0.25/mile (all inclusive)

Plan B: purchase the car for $30,000; keep the car for three years; sell the car after three years for $7200; pay $0.14 per mile for oil and gas; pay other costs of $500 per year

24. (*Time limit: one hour*) Two methods are being considered to meet strict air pollution control requirements over the next ten years. Method A uses equipment with a life of ten years. Method B uses equipment with a life of five years that will be replaced with new equipment with an additional life of five years. Capacities of the two methods are different, but operating costs do not depend on the throughput. Operation is 24 hours per day, 365 days per year. The effective annual interest rate for this evaluation is 7%.

	method A	method B	
	years 1–10	years 1–5	years 6–10
installation cost	$13,000	$6000	$7000
equipment cost	$10,000	$2000	$2200
operating cost per hour	$10.50	$8.00	$8.00
salvage value	$5000	$2000	$2000
capacity (tons/yr)	50	20	20
life	10 years	5 years	5 years

(a) What is the uniform annual cost per ton for each method? (b) Over what range of throughput (in units of tons/yr) does each method have the minimum cost?

25. (*Time limit: one hour*) A transit district has asked for your assistance in determining the proper fare for its bus system. An effective annual interest rate of 7% is to be used. The following additional information was compiled for your study.

cost per bus	$60,000
bus life	20 years
salvage value	$10,000
miles driven per year	37,440
number of passengers per year	80,000
operating cost	$1.00 per mile in the first year, increasing $0.10 per mile each year thereafter

(a) If the fare is to remain constant for the next 20 years, what is the break-even fare per passenger? (b) If the transit district decides to set the per-passenger fare at $0.35 for the first year, by what amount should the per-passenger fare go up each year thereafter such that the district can break even in 20 years? (c) If the transit district decides to set the per-passenger fare at $0.35 for the first year and the per-passenger fare goes up $0.05 each year thereafter, what additional governmental subsidy (per passenger) is needed for the district to break even in 20 years?

26. Make a recommendation to your client to accept one of the following alternatives. Use the present worth comparison method. (Initial costs are the same.)

Alternative A: a 25 year annuity paying $4800 at the end of each year, where the interest rate is a nominal 12% per annum

Alternative B: a 25 year annuity paying $1200 every quarter at 12% nominal annual interest

27. A firm has two alternatives for improvement of its existing production line. The data are as follows.

	alternative A	alternative B
initial installment cost	$1500	$2500
annual operating cost	$800	$650
service life	5 years	8 years
salvage value	0	0

Determine the best alternative using an interest rate of 15%.

28. Two mutually exclusive alternatives requiring different investments are being considered. The life of both alternatives is estimated at 20 years with no salvage values. The minimum rate of return that is considered acceptable is 4%. Which alternative is best?

	alternative A	alternative B
investment required	$70,000	$40,000
net income per year	$5620	$4075
rate of return on total investment	5%	8%

29. Compare the costs of two plant renovation schemes, A and B. Assume equal lives of 25 years, no salvage values, and interest at 25%. Make the comparison on the basis of (a) present worth, (b) capitalized cost, and (c) annual cost.

	alternative A	alternative B
first cost	$20,000	$25,000
annual expenditure	$3000	$2500

30. With interest at 8%, obtain the solutions to the following to the nearest dollar. (a) A machine costs $18,000 and has a salvage value of $2000. It has a useful life of 8 years. What is its book value at the end of 5 years using straight line depreciation? (b) Using data from part (a), find the depreciation in the first three years using the sinking fund method. (c) Repeat part (a) using double declining balance depreciation to find the first five years' depreciation.

31. A chemical pump motor unit is purchased for $14,000. The estimated life is 8 years, after which it will be sold for $1800. Find the depreciation in the first two years by the sum-of-the-years' digits method. Calculate the after-tax depreciation recovery using 15% interest with 52% income tax.

32. A soda ash plant has the water effluent from processing equipment treated in a large settling basin. The settling basin eventually discharges into a river that runs alongside the basin. Recently enacted environmental regulations require all rainfall on the plant to be diverted and treated in the settling basin. A heavy rainfall will cause the entire basin to overflow. An uncontrolled overflow will cause environmental damage and heavy fines. The construction of additional height on the existing basic walls is under consideration.

Data on the costs of construction and expected costs for environmental cleanup and fines are shown. Data on 50 typical winters have been collected. The soda ash plant management considers 12% to be their minimum rate of return, and it is felt that after 15 years the plant will be closed. The company wants to select the alternative that minimizes its total expected costs.

additional basin height (ft)	number of winters with basin overflow	expense for environmental clean up per year	construction cost
0	24	$550,000	0
5	14	$600,000	$600,000
10	8	$650,000	$710,000
15	3	$700,000	$900,000
20	1	$800,000	$1,000,000
	50		

33. A wood processing plant installed a waste gas scrubber at a cost of $30,000 to remove pollutants from the exhaust discharged into the atmosphere. The scrubber has no salvage value and will cost $18,700 to operate next year, with operating costs expected to increase at the rate of $1200 per year thereafter. When should the company consider replacing the scrubber? Money can be borrowed at 12%.

34. Two alternative piping schemes are being considered by a water treatment facility. On the basis of a 10-year life and an interest rate of 12%, determine the number of hours of operation for which the two installations will be equivalent.

	alternative A	alternative B
pipe diameter	4 in	6 in
head loss for required flow	48 ft	26 ft
size motor required	20 hp	7 hp
energy cost per hour of operation	$0.30	$0.10
cost of motor installed	$3600	$2800
cost of pipes and fittings	$3050	$5010
salvage value at end of 10 years	$200	$280

35. An 88% learning curve is used with an item whose first production time was 6 weeks. How long will it take to produce the fourth item? How long will it take to produce the sixth through fourteenth items?

36. (*Time limit: one hour*) A company is considering two alternatives, only one of which can be selected.

alternative	initial investment	salvage value	annual net profit	life
A	$120,000	$15,000	$57,000	5 yr
B	$170,000	$20,000	$67,000	5 yr

The net profit is after operating and maintenance costs, but before taxes. The company pays 45% of its year-end profit as income taxes. Use straight line depreciation. Do not use investment tax credit. Find the best alternative if the company's minimum attractive rate of return is 15%.

37. (*Time limit: one hour*) A company is considering the purchase of equipment to expand its capacity. The equipment cost is $300,000. The equipment is needed for 5 years, after which it will be sold for $50,000. The company's before-tax cash flow will be improved $90,000 annually by the purchase of the asset.

The corporate tax rate is 48%, and straight line depreciation will be used. The company will take an investment tax credit of 6.67%. What is the after-tax rate of return associated with this equipment purchase?

38. (*Time limit: one hour*) A 120-room hotel is purchased for $2,500,000. A 25-year loan is available for 12%. A study was conducted to determine the various occupancy rates.

occupancy	probability
65% full	0.40
70%	0.30
75%	0.20
80%	0.10

The operating costs of the hotel are as follows.

taxes and insurance	$20,000 annually
maintenance	$50,000 annually
operating	$200,000 annually

The life of the hotel is figured to be 25 years when operating 365 days per year. The salvage value after 25 years is $500,000.

Neglect tax credit and income taxes. Determine the average rate that should be charged per room per night to return 15% of the initial cost each year.

39. (*Time limit: one hour*) A company is insured for $3,500,000 against fire and the insurance rate is $0.69/1000. The insurance company will decrease the rate to $0.47/1000 if fire sprinklers are installed. The initial cost of the sprinklers is $7500. Annual costs are $200; additional taxes are $100 annually. The system life is 25 years. What is the rate of return on this investment?

40. (*Time limit: one hour*) Heat losses through the walls in an existing building cost a company $1,300,000 per year. This amount is considered excessive, and two alternatives are being evaluated. Neither of the alternatives will increase the life of the existing building beyond the current expected life of 6 years, and neither of the alternatives will produce a salvage value.

Alternative A: Do nothing. Continue with current losses.

Alternative B: Spend $2,000,000 immediately to upgrade the building and reduce the loss by 80%. Annual maintenance will cost $150,000.

Alternative C: Spend $1,200,000 immediately. Repeat the $1,200,000 expenditure 3 years from now. Heat loss the first year will be reduced 80%. Due to deterioration, the reduction will be 55% and 20% in the second and third years. (The pattern is repeated starting after the second expenditure.) There are no maintenance costs.

All energy and maintenance costs are considered expenses for tax purposes. The company's tax rate is 48%, and straight line depreciation is used. 15% is regarded as the effective annual interest rate. Evaluate each alternative on an after-tax basis, and recommend the best alternative.

41. (*Time limit: one hour*) You have been asked to determine if a 7-year-old machine should be replaced. Give a full explanation for your recommendation. Base your decision on a before-tax interest rate of 15%.

The existing machine is presumed to have a 10-year life. It has been depreciated on a straight line basis from its original value of $1,250,000 to a current book value of $620,000. Its ultimate salvage value was assumed to be $350,000 for purposes of depreciation. Its present salvage value is estimated at $400,000, and this is not expected to change over the next 3 years. The current operating costs are not expected to change from $200,000 per year.

A new machine costs $800,000, with operating costs of $40,000 the first year, and increasing by $30,000 each year thereafter. The new machine has an expected life of 10 years. The salvage value depends on the year the new machine is retired.

year retired	salvage
1	$600,000
2	$500,000
3	$450,000
4	$400,000
5	$350,000
6	$300,000
7	$250,000
8	$200,000
9	$150,000
10	$100,000

SOLUTIONS

1.

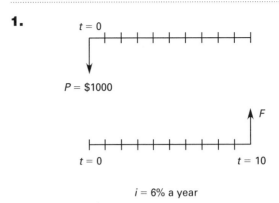

$i = 6\%$ a year

By the formula from Table 56.1,

$$F = P(1 + i)^n = (\$1000)(1 + 0.06)^{10} = \boxed{\$1790.85}$$

By the factor converting P to F, $(F/P, i, n) = 1.7908$ for $i = 6\%$ a year and $n = 10$ years.

$$F = P(F/P, 6\%, 10) = (\$1000)(1.7908) = \boxed{\$1790.80}$$

2.

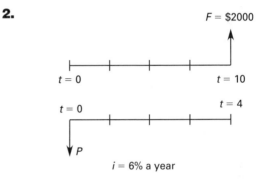

$i = 6\%$ a year

By the formula from Table 56.1,

$$P = \frac{F}{(1 + i)^n} = \frac{\$2000}{(1 + 0.06)^4} = \boxed{\$1584.19}$$

From the factor converting F to P, $(P/F, i, n) = 0.7921$ for $i = 6\%$ a year and $n = 4$ years.

$$P = F(P/F, 6\%, 4) = (\$2000)(0.7921) = \boxed{\$1584.20}$$

3.

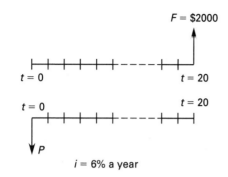

By the formula from Table 56.1,

$$P = \frac{F}{(1+i)^n} = \frac{\$2000}{(1+0.06)^{20}} = \boxed{\$623.61}$$

From the factor converting F to P, $(P/F, i, n) = 0.3118$ for $i = 6\%$ a year and $n = 20$ years.

$$P = F(P/F, 6\%, 20) = (\$2000)(0.3118) = \boxed{\$623.60}$$

4.

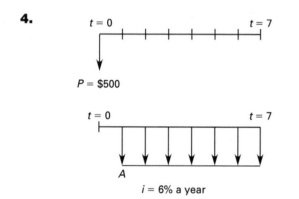

By the formula from Table 56.1,

$$A = P\left(\frac{i(1+i)^n}{(1+i)^n - 1}\right) = (\$500)\left(\frac{(0.06)(1+0.06)^7}{(1+0.06)^7 - 1}\right)$$

$$= \boxed{\$89.57}$$

By the factor converting P to A, $(A/P, i, n) = 0.17914$ for $i = 6\%$ a year and $n = 7$ years.

$$A = P(A/P, 6\%, 7) = (\$500)(0.17914) = \boxed{\$89.57}$$

5.

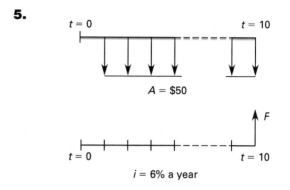

By the formula from Table 56.1,

$$F = A\left(\frac{(1+i)^n - 1}{i}\right) = (\$50)\left(\frac{(1+0.06)^{10} - 1}{0.06}\right)$$

$$= \boxed{\$659.04}$$

By the factor converting A to F, $(F/A, i, n) = 13.181$ for $i = 6\%$ a year and $n = 10$ years.

$$F = A(F/A, 6\%, 10) = (\$50)(13.181) = \boxed{\$659.05}$$

6.

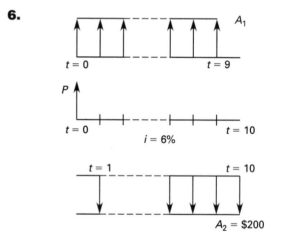

By the formula from Table 56.1, for each cash flow diagram,

$$P = A_1 + (A_1)\left(\frac{(1+0.06)^9 - 1}{(0.06)(1+0.06)^9}\right)$$

$$= A_2\left(\frac{(1+0.06)^{10} - 1}{(0.06)(1+0.06)^{10}}\right)$$

Therefore, for $A_2 = \$200$,

$$A_1 + (A_1)\left(\frac{(1+0.06)^9 - 1}{(0.06)(1+0.06)^9}\right)$$

$$= (\$200)\left(\frac{(1+0.06)^{10} - 1}{(0.06)(1+0.06)^{10}}\right)$$

$$7.80A_1 = \$1472.02$$

$$A_1 = \boxed{\$188.72}$$

By the factor converting A to P,

$$(P/A, 6\%, 9) = 6.8017$$

$$(P/A, 6\%, 10) = 7.3601$$

$$A_1 + A_1(6.8017) = (\$200)(7.3601)$$

$$7.8017A_1 = \$1472.02$$

$$A_1 = \frac{\$1472.02}{7.8017} = \boxed{\$188.68}$$

7.

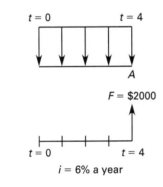

i = 6% a year

By the formula from Table 56.1,

$$F = A\left(\frac{(1+i)^n - 1}{i}\right)$$

Since the deposits start at the beginning of each year, $n = 4 + 1$, for a total of five deposits.

$$F = A\left(\frac{(1+i)^{n+1} - 1}{i}\right) = \$2000$$

$$= A\left(\frac{(1+0.06)^5 - 1}{0.06}\right)$$

$$\$2000 = 5.6371A$$

$$A = \frac{\$2000}{5.6371}$$

$$= \boxed{\$354.79}$$

By the factor converting P and A to F,

$$F = A\big((F/P, 6\%, 4) + (F/A, 6\%, 4)\big)$$

$$\$2000 = A(1.2625 + 4.3746)$$

$$A = \boxed{\$354.79}$$

8.

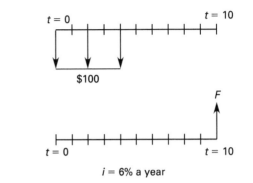

i = 6% a year

By the formula from Table 56.1, $F = P(1 + i)^n$. If each deposit is considered as P, each will accumulate interest for periods of 10, 8, and 6 years.

Therefore,

$$F = (\$100)(1+0.06)^{10} + (\$100)(1+0.06)^8$$

$$+ (\$100)(1+0.06)^6$$

$$= (\$100)(1.7908 + 1.5938 + 1.4185)$$

$$= (\$100)(4.8031)$$

$$= \boxed{\$480.31}$$

By the factor converting P to F,

$$(F/P, i, n) = 1.7908 \text{ for } i = 6\% \text{ and } n = 10$$

$$= 1.5938 \text{ for } i = 6\% \text{ and } n = 8$$

$$= 1.4185 \text{ for } i = 6\% \text{ and } n = 6$$

By summation,

$$F = (\$100)(1.7908 + 1.5938 + 1.4185)$$

$$= (\$100)(4.8031)$$

$$= \boxed{\$480.31}$$

9.

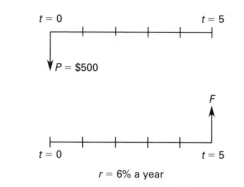

Since the deposit is compounded monthly, the effective interest rate should be calculated as shown by Eq. 56.54.

$$i = \left(1 + \frac{r}{k}\right)^k - 1 = \left(1 + \frac{0.06}{12}\right)^{12} - 1$$
$$= 0.061677 \quad (6.1677\%)$$

By the formula from Table 56.1,

$$F = P(1 + i)^n = (\$500)(1 + 0.061677)^5 = \boxed{\$674.42}$$

To use a table of factors, interpolation is required.

$i\%$	factor F/P
6	1.3382
6.1677	desired
7	1.4026

$$\Delta(F/P) = \left(\frac{6.1677\% - 6\%}{7\% - 6\%}\right)(1.4026 - 1.3382)$$
$$= 0.0108$$

Therefore,

$$F/P = 1.3382 + 0.0108 = 1.3490$$
$$F = P(F/P, 6.1677\%, 5) = (\$500)(1.3490)$$
$$= \boxed{\$674.50}$$

10.

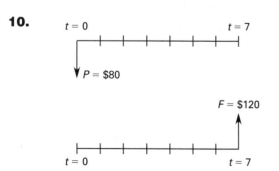

By the formula from Table 56.1,

$$F = P(1 + i)^n$$

Therefore,

$$(1 + i)^n = F/P$$
$$i = (F/P)^{1/n} - 1 = \left(\frac{\$120}{\$80}\right)^{1/7} - 1$$
$$= 0.059 \approx \boxed{6\%}$$

By the factor converting P to F,

$$F = P(F/P, i\%, 7)$$
$$(F/P, i\%, 7) = F/P = \frac{\$120}{\$80} = 1.5$$

By checking the interest tables,

$$(F/P, i\%, 7) = 1.4071 \text{ for } i = 5\%$$
$$= 1.5036 \text{ for } i = 6\%$$
$$= 1.6058 \text{ for } i = 7\%$$

Therefore, $i \approx \boxed{6\%.}$

11.

Annual cost of ownership, EUAC, can be obtained by the factors converting P to A and F to A.

$$P = \$17,000 + \$5000$$
$$= \$22,000$$
$$F = \$14,000 + \$2500$$
$$= \$16,500$$
$$EUAC = A + P(A/P, 6\%, 5) - F(A/F, 6\%, 5)$$
$$(A/P, 6\%, 5) = 0.23740$$
$$(A/F, 6\%, 5) = 0.17740$$
$$EUAC = \$2000 + (\$22,000)(0.23740)$$
$$- (\$16,500)(0.17740)$$
$$= \boxed{\$4295.70}$$

12.

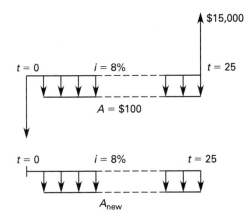

Consider the salvage value as a benefit lost (cost).

$$EUAC_{old} = \$500 + (\$22,000)(A/P, 8\%, 20)$$
$$- (\$10,000)(A/F, 8\%, 20)$$
$$(A/P, 8\%, 20) = 0.1019$$
$$(A/F, 8\%, 20) = 0.0219$$
$$EUAC_{old} = \$500 + (\$22,000)(0.1019)$$
$$- (\$10,000)(0.0219)$$
$$= \$2522.80$$

Similarly,

$$EUAC_{new} = \$100 + (\$40,000)(A/P, 8\%, 25)$$
$$- (\$15,000)(A/F, 8\%, 25)$$
$$(A/P, 8\%, 25) = 0.0937$$
$$(A/F, 8\%, 25) = 0.0137$$
$$EUAC_{new} = \$100 + (\$40,000)(0.0937)$$
$$- (\$15,000)(0.0137)$$
$$= \$3642.50$$

Therefore, the new bridge is going to be more costly.

$$\boxed{\text{The best alternative is to strengthen the old bridge.}}$$

13.

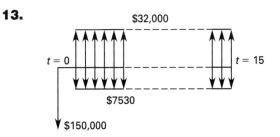

The annual depreciation is

$$D = \frac{C - S_n}{n} = \frac{\$150,000}{15}$$
$$= \$10,000/\text{year}$$

The taxable income is

$$\$32,000 - \$7530 - \$10,000 = \$14,470/\text{year}$$

Taxes paid are

$$(\$14,470)(0.48) = \$6945.60/\text{year}$$

The after-tax cash flow is

$$\$24{,}470 - \$6945.60 = \$17{,}524.40$$

The present worth of the alternate is zero when evaluated at its ROR.

$$0 = -\$150{,}000 + (\$17{,}524.40)(P/A, i\%, 15)$$

Therefore,

$$(P/A, i\%, 15) = \frac{\$150{,}000}{\$17{,}524.40} = 8.55949$$

By checking the tables, this factor matches $i = 8\%$.

$$\boxed{\text{ROR} = 8\%}$$

14. The conventional benefit/cost ratio is

$$B/C = \frac{B - D}{D}$$

(a) The benefit/cost ratio will be

$$B/C = \frac{\$1{,}500{,}000 - \$300{,}000}{\$1{,}000{,}000} = \boxed{1.2}$$

(b) The excess of benefits over cost is $\boxed{\$200{,}000.}$

15.

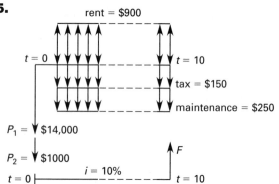

The annual rent is

$$(\$75)\left(12 \; \frac{\text{months}}{\text{year}}\right) = \$900$$

$$P = P_1 + P_2 = \$15{,}000$$

$$A_1 = -\$900$$

$$A_2 = \$250 + \$150 = \$400$$

By the factors converting P to F and A to F,

$$F = (\$15{,}000)(F/P, 10\%, 10)$$
$$+ (\$400)(F/A, 10\%, 10)$$
$$- (\$900)(F/A, 10\%, 10)$$

$$(F/P, 10\%, 10) = 2.5937$$
$$(F/A, 10\%, 10) = 15.937$$
$$F = (\$15{,}000)(2.5937) + (\$400)(15.937)$$
$$- (\$900)(15.937)$$
$$= \boxed{\$30{,}937}$$

16.

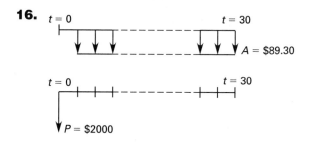

By the formula relating P to A,

$$P = A\left(\frac{(1 + i)^n - 1}{i(1 + i)^n}\right)$$

$$\frac{(1 + i)^{30} - 1}{i(1 + i)^{30}} = \frac{\$2000}{\$89.30} = 22.40$$

By trial and error,

$i\%$	$(1 + i)^{30}$	$\dfrac{(1 + i)^{30} - 1}{i(1 + i)^{30}}$
10	17.45	9.42
6	5.74	13.76
4	3.24	17.28
2	1.81	22.37

2% per month is close.

$$i = (1 + 0.02)^{12} - 1 = \boxed{0.2682 \quad (26.82\%)}$$

17. (a) Use the straight line method, Eq. 56.25.

$$D = \frac{C - S_n}{n}$$

Each year depreciation will remain the same.

$$D = \frac{\$500{,}000 - \$100{,}000}{25} = \boxed{\$16{,}000}$$

(b) Sum-of-the-years' digits (SOYD) can be calculated as shown by Eq. 56.28.

$$D_j = \frac{(C - S_n)(n - j + 1)}{T}$$

Use Eq. 56.27.

$$T = \tfrac{1}{2}n(n+1) = \left(\tfrac{1}{2}\right)(25)(25+1) = 325$$

$$D_1 = \frac{(\$500{,}000 - \$100{,}000)(25 - 1 + 1)}{325}$$

$$= \boxed{\$30{,}769}$$

$$D_2 = \frac{(\$500{,}000 - \$100{,}000)(25 - 2 + 1)}{325}$$

$$= \boxed{\$29{,}538}$$

$$D_3 = \frac{(\$500{,}000 - \$100{,}000)(25 - 3 + 1)}{325}$$

$$= \boxed{\$28{,}308}$$

(c) The double declining balance (DDB) method can be used. By Eq. 56.32,

$$D_j = dC(1 - d)^{j-1}$$

Use Eq. 56.31.

$$d = \tfrac{2}{n}$$

$$= \tfrac{2}{25}$$

$$D_1 = \left(\tfrac{2}{25}\right)(\$500{,}000)\left(1 - \tfrac{2}{25}\right)^0 = \boxed{\$40{,}000}$$

$$D_2 = \left(\tfrac{2}{25}\right)(\$500{,}000)\left(1 - \tfrac{2}{25}\right)^1 = \boxed{\$36{,}800}$$

$$D_3 = \left(\tfrac{2}{25}\right)(\$500{,}000)\left(1 - \tfrac{2}{25}\right)^2 = \boxed{\$33{,}856}$$

18.

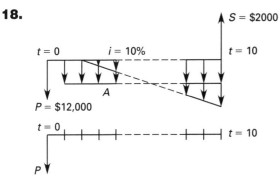

(a) $A = \$1000$ and $G = \$200$ for $t = n - 1 = 9$ years.

$$F = S = \$2000$$

$$P = \$12{,}000 + A(P/A, 10\%, 10) + G(P/G, 10\%, 10)$$
$$\quad - F(P/F, 10\%, 10)$$

$$= \$12{,}000 + (\$1000)(6.1446) + (\$200)(22.8913)$$
$$\quad - (\$2000)(0.3855)$$

$$= \boxed{\$21{,}952}$$

(b) $$A = (\$12{,}000)(A/P, 10\%, 10) + \$1000$$
$$\quad + (\$200)(A/G, 10\%, 10)$$
$$\quad - (\$2000)(A/F, 10\%, 10)$$

$$= (\$12{,}000)(0.1627) + \$1000 + (\$200)(3.7255)$$
$$\quad - (\$2000)(0.0627)$$

$$= \boxed{\$3572.10}$$

19. An increase in rock removal capacity can be achieved by a 20-year loan (investment). Different cases available can be compared by equivalent uniform annual cost (EUAC).

$$\text{EUAC} = \text{annual loan cost}$$
$$\quad + \text{expected annual damage}$$
$$= \text{cost}\,(A/P, 10\%, 20)$$
$$\quad + (\$25{,}000)(\text{probability})$$

$$(A/P, 10\%, 20) = 0.1175$$

A table can be prepared for different cases.

rock removal rate	cost ($)	annual loan cost ($)	expected annual damage ($)	EUAC ($)
7	0	0	3750	3750.00
8	15,000	1761.90	2500	4261.90
9	20,000	2349.20	1750	4099.20
10	30,000	3523.80	750	4273.80

$$\boxed{\text{It is cheapest to do nothing.}}$$

20. Calculate the cost of owning and operating for each year.

$$A_1 = (\$10{,}000)(A/P, 20\%, 1) + \$2000$$
$$- (\$8000)(A/F, 20\%, 1)$$

$$(A/P, 20\%, 1) = 1.2$$

$$(A/F, 20\%, 1) = 1.0$$

$$A_1 = (\$10{,}000)(1.2) + \$2000 - (\$8000)(1.0)$$
$$= \$6000$$

$$A_2 = (\$10{,}000)(A/P, 20\%, 2) + \$2000$$
$$+ (\$1000)(A/G, 20\%, 2)$$
$$- (\$7000)(A/F, 20\%, 2)$$

$$(A/P, 20\%, 2) = 0.6545$$

$$(A/G, 20\%, 2) = 0.4545$$

$$(A/F, 20\%, 2) = 0.4545$$

$$A_2 = (\$10{,}000)(0.6545) + \$2000$$
$$+ (\$1000)(0.4545) - (\$7000)(0.4545)$$
$$= \$5818$$

$$A_3 = (\$10{,}000)(A/P, 20\%, 3) + \$2000$$
$$+ (\$1000)(A/G, 20\%, 3)$$
$$- (\$6000)(A/F, 20\%, 3)$$

$$(A/P, 20\%, 3) = 0.4747$$

$$(A/G, 20\%, 3) = 0.8791$$

$$(A/F, 20\%, 3) = 0.2747$$

$$A_3 = (\$10{,}000)(0.4747) + \$2000$$
$$+ (\$1000)(0.8791)$$
$$- (\$6000)(0.2747)$$
$$= \$5977.90$$

$$A_4 = (\$10{,}000)(A/P, 20\%, 4)$$
$$+ \$2000 + (\$1000)(A/G, 20\%, 4)$$
$$- (\$5000)(A/F, 20\%, 4)$$

$$(A/P, 20\%, 4) = 0.3863$$

$$(A/G, 20\%, 4) = 1.2762$$

$$(A/F, 20\%, 4) = 0.1863$$

$$A_4 = (\$10{,}000)(0.3863) + \$2000$$
$$+ (\$1000)(1.2762) - (\$5000)(0.1863)$$
$$= \$6207.70$$

$$A_5 = (\$10{,}000)(A/P, 20\%, 5) + \$2000$$
$$+ (\$1000)(A/G, 20\%, 5)$$
$$- (\$4000)(A/F, 20\%, 5)$$

$$(A/P, 20\%, 5) = 0.3344$$

$$(A/\dot{G}, 20\%, 5) = 1.6405$$

$$(A/F, 20\%, 5) = 0.1344$$

$$A_5 = (\$10{,}000)(0.3344) + \$2000$$
$$+ (\$1000)(1.6405) - (\$4000)(0.1344)$$
$$= \$6446.90$$

(a) Since the annual owning and operating cost is smallest after two years of operation, it is advantageous to sell the mechanism after the second year.

$$\boxed{\text{The economic life is two years.}}$$

(b) After four years of operation, the owning and operating cost of the mechanism for one more year will be

$$A = \$6000 + (\$5000)(1 + i) - \$4000$$

$$i = 0.2 \quad (20\%)$$

$$A = \$6000 + (\$5000)(1.2) - \$4000$$

$$= \boxed{\$8000}$$

21. To find out if the reimbursement is adequate, calculate the business-related expense.

Charge the company for business travel.

$$\begin{aligned}
\text{insurance:} \quad & \$3000 - \$2000 = \$1000 \\
\text{maintenance:} \quad & \$2000 - \$1500 = \$500 \\
\text{drop in salvage value:} \quad & \$10{,}000 - \$5000 = \$5000
\end{aligned}$$

The annual portion of the drop in salvage value is

$$A = (\$5000)(A/F, 10\%, 5)$$

$$(A/F, 10\%, 5) = 0.1638$$

$$A = (\$5000)(0.1638) = \$819/\text{year}$$

(a) The annual cost of gas is

$$\left(\frac{50{,}000 \text{ mi}}{15 \frac{\text{mi}}{\text{gal}}}\right)\left(\frac{\$1.50}{\text{gal}}\right) = \$5000$$

$$\text{EUAC per mile} = \frac{\$1000 + \$500 + \$819 + \$5000}{50{,}000 \text{ mi}}$$

$$= \boxed{\$0.14638/\text{mi}}$$

Since the reimbursement per mile was $0.30 and since $0.30 > $0.14638, the reimbursement is adequate.

(b) Next, determine (with reimbursement) how many miles the car must be driven to break even.

If the car is driven M miles per year,

$$\left(\frac{\$0.30}{1 \text{ mi}}\right) M = (\$50,000)(A/P, 10\%, 5) + \$2500$$
$$+ \$2000 - (\$8000)(A/F, 10\%, 5)$$
$$+ \left(\frac{M}{15 \frac{\text{mi}}{\text{gal}}}\right)(\$1.50)$$

$$(A/P, 10\%, 5) = 0.2638$$
$$(A/F, 10\%, 5) = 0.1638$$
$$0.3M = (\$50,000)(0.2638) + \$2500 + \$2000$$
$$- (\$8000)(0.1638) + 0.1M$$
$$0.2M = \$16,379.60$$
$$M = \frac{\$16,379.60}{0.2 \frac{\$}{\text{mi}}} = \boxed{81,898 \text{ mi}}$$

22.

$$P_{\text{A}} = \$80,000 + (\$1000)(P/A, 10\%, 5)$$
$$+ (\$1500)(P/A, 10\%, 5)(P/F, 10\%, 5)$$
$$+ (\$2000)(P/A, 10\%, 10)(P/F, 10\%, 10)$$
$$+ (\$5000)(P/F, 10\%, 10)$$
$$- (\$7000)(P/F, 10\%, 20)$$

$$(P/A, 10\%, 5) = 3.7908$$
$$(P/F, 10\%, 5) = 0.6209$$
$$(P/A, 10\%, 10) = 6.1446$$
$$(P/F, 10\%, 10) = 0.3855$$
$$(P/F, 10\%, 20) = 0.1486$$
$$P_{\text{A}} = \$80,000 + (\$1000)(3.7908)$$
$$+ (\$1500)(3.7908)(0.6209)$$
$$+ (\$2000)(6.1446)(0.3855)$$
$$+ (\$5000)(0.3855) - (\$7000)(0.1486)$$
$$= \$92,946.15$$

Since the lives are different, compare by EUAC.

$$\text{EUAC(A)} = (\$92,946.14)(A/P, 10\%, 20)$$
$$= (\$92,946.14)(0.1175) = \$10,921$$

Similarly, evaluate alternative B.

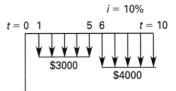

$$P_{\text{B}} = \$35,000 + (\$3000)(P/A, 10\%, 5)$$
$$+ (\$4000)(P/A, 10\%, 5)(P/F, 10\%, 5)$$

$$(P/A, 10\%, 5) = 3.7908$$
$$(P/F, 10\%, 5) = 0.6209$$
$$P_{\text{B}} = \$35,000 + (\$3000)(3.7908)$$
$$+ (\$4000)(3.7908)(0.6209)$$
$$= \$55,787.23$$
$$\text{EUAC(B)} = (\$55,787.23)(A/P, 10\%, 10)$$
$$= (\$55,787.23)(0.1627) = \$9077$$

Since EUAC(B) < EUAC(A),

$$\boxed{\text{Alternative B is economically superior.}}$$

23. For both cases, if the annual cost is compared with a total annual mileage of M,

$$A_{\text{A}} = \$0.25M$$
$$A_{\text{B}} = (\$30,000)(A/P, 10\%, 3) + \$0.14M$$
$$+ \$500 - (\$7200)(A/F, 10\%, 3)$$

$$(A/P, 10\%, 3) = 0.4021$$
$$(A/F, 10\%, 3) = 0.3021$$
$$A_{\text{B}} = (\$30,000)(0.4021) + \$0.14M + \$500$$
$$- (\$7200)(0.3021)$$
$$= \$12,063 + \$0.14M$$
$$+ \$500 - \$2175.12$$
$$= \$10,387.88 + \$0.14M$$

For an equal annual cost $A_{\text{A}} = A_{\text{B}}$,

$$\$0.25M = \$10,387.88 + \$0.14M$$

An annual mileage would be $M = 94{,}435$ mi.

For an annual mileage less than that, $A_A < A_B$.

> Plan A is economically superior until that mileage is exceeded.

24. (a) Method A:

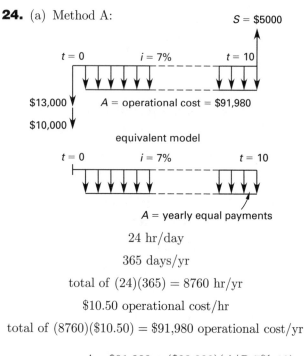

24 hr/day

365 days/yr

total of $(24)(365) = 8760$ hr/yr

$10.50 operational cost/hr

total of $(8760)(\$10.50) = \$91{,}980$ operational cost/yr

$$A = \$91{,}980 + (\$23{,}000)(A/P, 7\%, 10)$$
$$- (\$5000)(A/F, 7\%, 10)$$

$$(A/P, 7\%, 10) = 0.1424$$
$$(A/F, 7\%, 10) = 0.0724$$

$$A = \$91{,}980 + (\$23{,}000)(0.1424)$$
$$- (\$5000)(0.0724)$$
$$= \$94{,}893.20/\text{yr}$$

Therefore, the uniform annual cost per ton each year will be

$$\frac{\$94{,}893.20}{50 \text{ ton}} = \boxed{\$1897.86}$$

Method B:

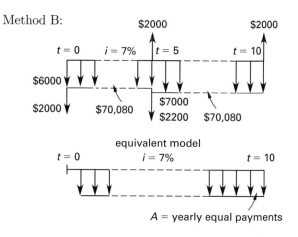

8760 hr/yr

$8 operational cost/hr

total of $70,080 operational cost/yr

$$A = \$70{,}080 + (\$6000 + \$2000)(A/P, 7\%, 10)$$
$$+ (\$7000 + \$2200 - \$2000)(P/F, 7\%, 5)$$
$$\times (A/P, 7\%, 10)$$
$$- (\$2000)(A/F, 7\%, 10)$$

$$(A/P, 7\%, 10) = 0.1424$$
$$(A/F, 7\%, 10) = 0.0724$$
$$(P/F, 7\%, 5) = 0.7130$$

$$A = \$70{,}080 + (\$8000)(0.1424)$$
$$+ (\$7200)(0.7130)(0.1424)$$
$$- (\$2000)(0.0724)$$
$$= \$71{,}805.42/\text{yr}$$

Therefore, the uniform annual cost per ton each year will be

$$\frac{\$71{,}805.42}{20 \text{ ton}} = \boxed{\$3590.27}$$

(b)

tons/yr	cost of using A		cost of using B		cheapest
0–20	$94,893	(1x)	$71,805	(1x)	B
20–40	$94,893	(1x)	$143,610	(2x)	A
40–50	$94,893	(1x)	$215,415	(3x)	A
50–60	$189,786	(2x)	$215,415	(3x)	A
60–80	$189,786	(2x)	$287,220	(4x)	A

25.

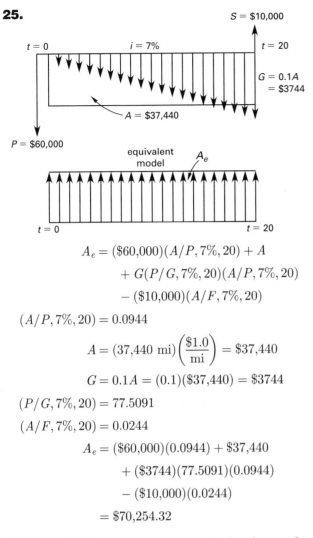

$$A_e = (\$60,000)(A/P, 7\%, 20) + A$$
$$+ G(P/G, 7\%, 20)(A/P, 7\%, 20)$$
$$- (\$10,000)(A/F, 7\%, 20)$$

$$(A/P, 7\%, 20) = 0.0944$$

$$A = (37{,}440 \text{ mi})\left(\frac{\$1.0}{\text{mi}}\right) = \$37{,}440$$

$$G = 0.1A = (0.1)(\$37{,}440) = \$3744$$

$$(P/G, 7\%, 20) = 77.5091$$

$$(A/F, 7\%, 20) = 0.0244$$

$$A_e = (\$60,000)(0.0944) + \$37{,}440$$
$$+ (\$3744)(77.5091)(0.0944)$$
$$- (\$10,000)(0.0244)$$
$$= \$70{,}254.32$$

(a) With 80,000 passengers a year, the break-even fare per passenger would be

$$\text{fare} = \frac{A_e}{80{,}000} = \frac{\$70{,}254.32}{80{,}000} = \boxed{\$0.878/\text{passenger}}$$

(b)
$$\$0.878 = \$0.35 + G(A/G, 7\%, 20)$$

$$G = \frac{\$0.878 - \$0.35}{7.3163}$$

$$= \boxed{\$0.072 \text{ increase per year}}$$

(c) As in part (b), the subsidy should be

$$\text{subsidy} = \text{cost} - \text{revenue}$$

$$P = \$0.878 - \big(\$0.35 + (\$0.05)(A/G, 7\%, 20)\big)$$

$$= \$0.878 - \big(\$0.35 + (\$0.05)(7.3163)\big)$$

$$= \boxed{\$0.162}$$

26.
$$P(A) = (\$4800)(P/A, 12\%, 25)$$
$$= (\$4800)(7.8431)$$
$$= \$37{,}646.88$$

$$(4 \text{ quarters})(25 \text{ years}) = 100 \text{ compounding periods}$$

$$P(B) = (\$1200)(P/A, 3\%, 100)$$
$$= (\$1200)(31.5989)$$
$$= \$37{,}918.68$$

$$\boxed{\text{Alternative B is economically superior.}}$$

27.
$$\text{EUAC(A)} = (\$1500)(A/P, 15\%, 5) + \$800$$
$$= (\$1500)(0.2983) + \$800$$
$$= \$1247.45$$
$$\text{EUAC(B)} = (\$2500)(A/P, 15\%, 8) + \$650$$
$$= (\$2500)(0.2229) + \$650$$
$$= \$1207.25$$

$$\boxed{\text{Alternative B is economically superior.}}$$

28. The data given imply that both investments return 4% or more. However, the increased investment of $30,000 may not be cost effective. Do an incremental analysis.

$$\text{incremental cost} = \$70{,}000 - \$40{,}000 = \$30{,}000$$

$$\text{incremental income} = \$5620 - \$4075 = \$1545$$

$$0 = -\$30{,}000 + (\$1545)(P/A, i\%, 20)$$

$$(P/A, i\%, 20) = 19.417$$

$$i \approx 0.25\% < 4\%$$

$$\boxed{\text{Alternative B is economically superior.}}$$

(The same conclusion could be reached by taking the present worths of both alternatives at 4%.)

29. (a) $P(A) = (-\$3000)(P/A, 25\%, 25) - \$20{,}000$

$\qquad = (-\$3000)(3.9849) - \$20{,}000$

$\qquad = -\$31{,}954.70$

$P(B) = (-\$2500)(3.9849) - \$25{,}000$

$\qquad = -\$34{,}962.25$

$\boxed{\text{Alternative A is better.}}$

(b) $CC(A) = \$20{,}000 + \dfrac{\$3000}{0.25} = \$32{,}000$

$CC(B) = \$25{,}000 + \dfrac{\$2500}{0.25} = \$35{,}000$

$\boxed{\text{Alternative A is better.}}$

(c) $EUAC(A) = (\$20{,}000)(A/P, 25\%, 25) + \$3000$

$\qquad = (\$20{,}000)(0.2509) + \$3000$

$\qquad = \$8018.00$

$EUAC(B) = (\$25{,}000)(0.2509) + \$2500$

$\qquad = \$8772.50$

$\boxed{\text{Alternative A is better.}}$

30. (a) $BV = \$18{,}000 - (5)\left(\dfrac{\$18{,}000 - \$2000}{8}\right)$

$\qquad = \boxed{\$8000}$

(b) With the sinking fund method, the basis is

$(\$18{,}000 - \$2000)(A/F, 8\%, 8)$

$\qquad = (\$18{,}000 - \$2000)(0.0940)$

$\qquad = \$1504$

$D_1 = (\$1504)(1.000) = \boxed{\$1504}$

$D_2 = (\$1504)(1.0800) = \boxed{\$1624}$

$D_3 = (\$1504)(1.0800)^2 = (\$1504)(1.1664)$

$\qquad = \boxed{\$1754}$

(c) $D_1 = \left(\tfrac{2}{8}\right)(\$18{,}000) = \boxed{\$4500}$

$D_2 = \left(\tfrac{2}{8}\right)(\$18{,}000 - \$4500) = \boxed{\$3375}$

$D_3 = \left(\tfrac{2}{8}\right)(\$18{,}000 - \$4500 - \$3375) = \boxed{\$2531}$

$D_4 = \left(\tfrac{2}{8}\right)(\$18{,}000 - \$4500 - \$3375 - \$2531)$

$\qquad = \boxed{\$1898}$

$D_5 = \left(\tfrac{2}{8}\right)(\$18{,}000 - \$4500 - \$3375$

$\qquad - \$2531 - \$1898)$

$\qquad = \boxed{\$1424}$

$BV = \$18{,}000 - \$4500 - \$3375 - \$2531$

$\qquad - \$1898 - \$1424$

$\qquad = \boxed{\$4272}$

31. $T = \left(\tfrac{1}{2}\right)(8)(9) = 36$

$D_1 = \left(\tfrac{8}{36}\right)(\$14{,}000 - \$1800) = \$2711$

$\Delta D = \left(\tfrac{1}{36}\right)(\$14{,}000 - \$1800) = \$339$

$D_2 = \$2711 - \$339 = \$2372$

$DR = (0.52)(\$2711)(P/A, 15\%, 8)$

$\qquad - (0.52)(\$339)(P/G, 15\%, 8)$

$\qquad = (0.52)(\$2711)(4.4873)$

$\qquad - (0.52)(\$339)(12.4807)$

$\qquad = \boxed{\$4125.74}$

32. $EUAC_{5\text{ ft}} = (\$600{,}000)(A/P, 12\%, 15)$

$\qquad + \left(\tfrac{14}{50}\right)(\$600{,}000) + \left(\tfrac{8}{50}\right)(\$650{,}000)$

$\qquad + \left(\tfrac{3}{50}\right)(\$700{,}000) + \left(\tfrac{1}{50}\right)(\$800{,}000)$

$\qquad = \$418{,}080$

$EUAC_{10\text{ ft}} = (\$710{,}000)(A/P, 12\%, 15)$

$\qquad + \left(\tfrac{8}{50}\right)(\$650{,}000) + \left(\tfrac{3}{50}\right)(\$700{,}000)$

$\qquad + \left(\tfrac{1}{50}\right)(\$800{,}000)$

$\qquad = \$266{,}228$

$EUAC_{15\text{ ft}} = (\$900{,}000)(A/P, 12\%, 15)$

$\qquad + \left(\tfrac{3}{50}\right)(\$700{,}000) + \left(\tfrac{1}{50}\right)(\$800{,}000)$

$\qquad = \$190{,}120$

$EUAC_{20\text{ ft}} = (\$1{,}000{,}000)(A/P, 12\%, 15)$

$\qquad + \left(\tfrac{1}{50}\right)(\$800{,}000)$

$\qquad = \$162{,}800$

$\boxed{\text{Build to 20 ft.}}$

33. Assume replacement after 1 year.

$$\text{EUAC}(1) = (\$30{,}000)(A/P, 12\%, 1) + \$18{,}700$$
$$= (\$30{,}000)(1.12) + \$18{,}700 = \$52{,}300$$

Assume replacement after 2 years.

$$\text{EUAC}(2) = (\$30{,}000)(A/P, 12\%, 2)$$
$$+ \$18{,}700 + (\$1200)(A/G, 12\%, 2)$$
$$= (\$30{,}000)(0.5917) + \$18{,}700$$
$$+ (\$1200)(0.4717) = \$37{,}017$$

Assume replacement after 3 years.

$$\text{EUAC}(3) = (\$30{,}000)(A/P, 12\%, 3)$$
$$+ \$18{,}700 + (\$1200)(A/G, 12\%, 3)$$
$$= (\$30{,}000)(0.4163) + \$18{,}700$$
$$+ (\$1200)(0.9246)$$
$$= \$32{,}299$$

Similarly, calculate to obtain the numbers in the following table.

years in service	EUAC
1	\$52,300
2	\$37,017
3	\$32,299
4	\$30,207
5	\$29,152
6	\$28,602
7	\$28,335
8	\$28,234
9	\$28,240
10	\$28,312

Replace after 8 yr.

34. Assume the head and horsepower data are already reflected in the hourly operating costs.

Let $N =$ no. of hours operated each year.

$$\text{EUAC}(A) = (\$3600 + \$3050)(A/P, 12\%, 10)$$
$$- (\$200)(A/F, 12\%, 10) + 0.30N$$
$$= (\$6650)(0.1770) - (\$200)(0.0570) + 0.30N$$
$$= 1165.65 + 0.30N$$
$$\text{EUAC}(B) = (\$2800 + \$5010)(A/P, 12\%, 10)$$
$$+ (\$280)(A/F, 12\%, 10) + 0.10N$$
$$= (\$7810)(0.1770) - (\$280)(0.0570) + 0.10N$$
$$= 1366.41 + 0.10N$$

$$\text{EUAC}(A) = \text{EUAC}(B)$$
$$1165.65 + 0.30N = 1366.41 + 0.10N$$
$$N = \boxed{1003.8 \text{ hr}}$$

35. (a) From Eq. 56.89,

$$\frac{T_2}{T_1} = 0.88 = 2^{-b}$$
$$\log 0.88 = -b \log 2$$
$$-0.0555 = -(0.3010)b$$
$$b = 0.1843$$

$$T_4 = (6)(4)^{-0.1843} = \boxed{4.65 \text{ wk}}$$

(b) From Eq. 56.90,

$$T_{6-14} = \left(\frac{6}{1 - 0.1843}\right)$$
$$\times \left(\left(14 + \tfrac{1}{2}\right)^{1 - 0.1843} - \left(6 - \tfrac{1}{2}\right)^{1 - 0.1843}\right)$$
$$= \left(\frac{6}{0.8157}\right)(8.857 - 4.017)$$
$$= \boxed{35.6 \text{ wk}}$$

36. First check that both alternatives have an ROR greater than the MARR. Work in thousands of dollars. Evaluate alternative A.

$$P(A) = -\$120 + (\$15)(P/F, i\%, 5)$$
$$+ (\$57)(P/A, i\%, 5)(1 - 0.45)$$
$$+ \left(\frac{\$120 - \$15}{5}\right)(P/A, i\%, 5)(0.45)$$
$$= -\$120 + (\$15)(P/F, i\%, 5)$$
$$+ (\$40.8)(P/A, i\%, 5)$$

Try 15%.

$$P(A) = -\$120 + (\$15)(0.4972) + (\$40.8)(3.3522)$$
$$= \$24.23$$

Try 25%.

$$P(A) = -\$120 + (\$15)(0.3277) + (\$40.8)(2.6893)$$
$$= -\$5.36$$

Since $P(A)$ goes through 0,

$$(\text{ROR})_A > \text{MARR} = 15\%$$

Next, evaluate alternative B.

$$P(B) = -\$170 + (\$20)(P/F, i\%, 5)$$
$$+ (\$67)(P/A, i\%, 5)(1 - 0.45)$$
$$+ \left(\frac{\$170 - \$20}{5}\right)(P/A, i\%, 5)(0.45)$$
$$= -\$170 + (\$20)(P/F, i\%, 5)$$
$$+ (\$50.35)(P/A, i\%, 5)$$

Try 15%.

$$P(B) = -\$170 + (\$20)(0.4972) + (\$50.35)(3.352)$$
$$= \$8.72$$

Since $P(B) > 0$ and will decrease as i increases,

$$(ROR)_B > 15\%$$

ROR > MARR for both alternatives.

Do an incremental analysis to see if it is worthwhile to invest the extra $\$170 - \$120 = \$50$.

$$P(B - A) = -\$50 + (\$20 - \$15)(P/F, i\%, 5)$$
$$+ (\$50.35 - \$40.8)(P/A, i\%, 5)$$

Try 15%.

$$P(B - A) = -\$50 + (\$5)(0.4972)$$
$$+ (\$9.55)(3.3522)$$
$$= -\$15.50$$

Since $P(B - A) < 0$ and would become more negative as i increases, the ROR of the added investment is greater than 15%.

Alternative A is superior.

37. Use the year-end convention with the tax credit. The purchase is made at $t = 0$. However, the credit is received at $t = 1$ and must be multiplied by $(P/F, i\%, 1)$.

(0.0667 is actually $^2/_3$ of 10%.)

$$P = -\$300,000 + (0.0667)(\$300,000)(P/F, i\%, 1)$$
$$+ (\$90,000)(P/A, i\%, 5)(1 - 0.48)$$
$$+ \left(\frac{\$300,000 - \$50,000}{5}\right)(P/A, i\%, 5)(0.48)$$
$$+ (\$50,000)(P/F, i\%, 5)$$
$$= -\$300,000 + (\$20,000)(P/F, i\%, 1)$$
$$+ (\$46,800)(P/A, i\%, 5)$$
$$+ (\$24,000)(P/A, i\%, 5)$$
$$+ (\$50,000)(P/F, i\%, 5)$$

By trial and error,

i	P
10%	$17,616
15%	−$20,412
12%	$1448
13%	−$6142
12 $\frac{1}{4}$%	−$479

i is between 12% and $12^1/_4$%.

38. Assume loan payments are made at the end of each year. Find the annual payment.

$$\text{payment} = (\$2,500,000)(A/P, 12\%, 25)$$
$$= (\$2,500,000)(0.1275)$$
$$= \$318,750$$
$$\text{distributed profit} = (0.15)(\$2,500,000)$$
$$= \$375,000$$

After paying all expenses and distributing the 15% profit, the remainder should be 0.

$$0 = \text{EUAC} = \$20,000 + \$50,000 + \$200,000$$
$$+ \$375,000 + \$318,750 - \text{annual receipts}$$
$$- (\$500,000)(A/F, 15\%, 25)$$
$$= \$963,750 - \text{annual receipts}$$
$$- (\$500,000)(0.0047)$$

This calculation assumes $i = 15\%$, which equals the desired return. However, this assumption only affects the salvage calculation, and since the number is so small, the analysis is not sensitive to the assumption.

$$\text{annual receipts} = \$961,400$$

The average daily receipts are

$$\frac{\$961,400}{365} = \$2634$$

Use the expected value approach. The average occupancy is

$$(0.40)(0.65) + (0.30)(0.70) + (0.20)(0.75)$$
$$+ (0.10)(0.80) = 0.70$$

The average number of rooms occupied each night is

$$(0.70)(120 \text{ rooms}) = 84 \text{ rooms}$$

The minimum required average daily rate per room is

$$\frac{\$2634}{84} = \boxed{\$31.36}$$

39.
$$\begin{aligned}\text{annual} \atop \text{savings}\end{aligned} = \left(\frac{0.69 - 0.47}{1000}\right)(\$3,500,000) = \$770$$
$$P = -\$7500 + (\$770 - \$200 - \$100)$$
$$\times (P/A, i\%, 25) = 0$$
$$(P/A, i\%, 25) = 15.957$$

Searching the tables and interpolating,

$$i \approx \boxed{3.75\%}$$

40. Evaluate alternative A, working in millions of dollars.

$$\begin{aligned}P(A) &= -(\$1.3)(1 - 0.48)(P/A, 15\%, 6) \\ &= (\$1.3)(0.52)(3.7845) \\ &= -\$2.56 \quad \text{[millions]}\end{aligned}$$

Since this is an after-tax analysis and since the salvage value was mentioned, assume that the improvements can be depreciated.

Use straight line depreciation.

Evaluate alternative B.

$$D_j = \tfrac{2}{6} = 0.333$$
$$\begin{aligned}P(B) &= -\$2 - (\$0.20)(\$1.3)(1 - 0.48)(P/A, 15\%, 6) \\ &\quad - (\$0.15)(1 - 0.48)(P/A, 15\%, 6) \\ &\quad + (\$0.333)(0.48)(P/A, 15\%, 6) \\ &= -\$2 - (\$0.20)(\$1.3)(0.52)(3.7845) \\ &\quad - (\$0.15)(0.52)(3.7845) \\ &\quad + (\$0.333)(0.48)(3.7845) \\ &= -\$2.206 \quad \text{[millions]}\end{aligned}$$

Next, evaluate alternative C.

$$D_j = \frac{1.2}{3} = 0.4$$
$$\begin{aligned}P(C) &= -(\$1.2)\big(1 + (P/F, 15\%, 3)\big) \\ &\quad - (\$0.20)(\$1.3)(1 - 0.48) \\ &\quad \times \big((P/F, 15\%, 1) + (P/F, 15\%, 4)\big) \\ &\quad - (\$0.45)(\$1.3)(1 - 0.48) \\ &\quad \times \big((P/F, 15\%, 2) + (P/F, 15\%, 5)\big) \\ &\quad - (\$0.80)(\$1.3)(1 - 0.48) \\ &\quad \times \big((P/F, 15\%, 3) + (P/F, 15\%, 6)\big) \\ &\quad + (\$0.4)(\$0.48)(P/A, 15\%, 6) \\ &= -(\$1.2)(1.6575) \\ &\quad - (\$0.20)(\$1.3)(0.52)(0.8696 + 0.5718) \\ &\quad - (\$0.45)(\$1.3)(0.52)(0.7561 + 0.4972) \\ &\quad - (\$0.80)(\$1.3)(0.52)(0.6575 + 0.4323) \\ &\quad + (\$0.4)(0.48)(3.7845) \\ &= -\$2.428 \quad \text{[millions]}\end{aligned}$$

$$\boxed{\text{Alternative B is superior.}}$$

41. This is a replacement study. Since production capacity and efficiency are not a problem with the defender, the only question is when to bring in the challenger.

Since this is a before-tax problem, depreciation is not a factor, nor is book value.

The cost of keeping the defender one more year is

$$\begin{aligned}\text{EUAC(defender)} &= \$200,000 + (0.15)(\$400,000) \\ &= \$260,000\end{aligned}$$

For the challenger,

$$\text{EUAC(challenger)}$$
$$= (\$800{,}000)(A/P, 15\%, 10) + \$40{,}000$$
$$+ (\$30{,}000)(A/G, 15\%, 10)$$
$$- (\$100{,}000)(A/F, 15\%, 10)$$
$$= (\$800{,}000)(0.1993) + \$40{,}000$$
$$+ (\$30{,}000)(3.3832)$$
$$- (\$100{,}000)(0.0493)$$
$$= \$296{,}006$$

Since the defender is cheaper, keep it. The same analysis next year will give identical answers. Therefore, keep the defender for the next 3 years, at which time the decision to buy the challenger will be automatic.

Having determined that it is less expensive to keep the defender than to maintain the challenger for 10 years, determine whether the challenger is less expensive if retired before 10 years.

If retired in 9 years,

$$\text{EUAC(challenger)} = (\$800{,}000)(A/P, 15\%, 9) + \$40{,}000$$
$$+ (\$30{,}000)(A/G, 15\%, 9)$$
$$- (\$150{,}000)(A/F, 15\%, 9)$$
$$= (\$800{,}000)(0.2096)$$
$$+ \$40{,}000 + (\$30{,}000)(3.0922)$$
$$- (\$150{,}000)(0.0596)$$
$$= \$291{,}506$$

Similar calculations yield the following results for all the retirement dates.

n	EUAC
10	\$296,000
9	\$291,506
8	\$287,179
7	\$283,214
6	\$280,016
5	\$278,419
4	\$279,909
3	\$288,013
2	\$313,483
1	\$360,000

Since none of these equivalent uniform annual costs is less than that of the defender, it is not economical to buy and keep the challenger for any length of time.

Keep the defender.

57 Engineering Law

PRACTICE PROBLEMS

1. List the different forms of company ownership. What are the advantages and disadvantages of each?

2. Define the requirements for a contract to be enforceable.

3. What standard features should a written contract include?

4. Describe the ways a consulting fee can be structured.

5. What is a retainer fee?

6. Which of the following organizations is NOT a contributor to the standard design and construction contract documents developed by the Engineers Joint Contract Documents Committee (EJCDC)?

(A) National Society of Professional Engineers

(B) Construction Specifications Institute

(C) Associated General Contractors of America

(D) American Institute of Architects

7. To be affected by the Fair Labor Standards Act (FLSA) and required to pay minimum wage, construction firms working on bridges and highways must generally have

(A) 1 or more employees

(B) 2 or more employees and annual gross billings of $500,000

(C) 10 or more employees and be working on federally funded projects

(D) 50 or more employees and have been in business for longer than 6 months

8. A double-breasted design firm

(A) has errors and omissions as well as general liability insurance coverage

(B) is licensed to practice in both engineering and architecture

(C) serves both union and nonunion clients

(D) performs post-construction certification for projects it did not design

9. The phrase "without expressed authority" means which of the following when used in regards to partnerships of design professionals?

(A) Each full member of a partnership is a general agent of the partnership and has complete authority to make binding commitments, enter into contracts, and otherwise act for the partners within the scope of the business.

(B) The partnership may act in a manner that it considers best for the client, even though the client has not been consulted.

(C) Only plans, specifications, and documents that have been signed and stamped (sealed) by the authority of the licensed engineer may be relied upon.

(D) Only officers to the partnership may obligate the partnership.

10. A limited partnership has 1 managing general partner, 2 general partners, 1 silent partner, and 3 limited partners. If all partners cast a single vote when deciding on an issue, how many votes will be cast?

(A) 1

(B) 3

(C) 4

(D) 7

11. Which of the following are characteristics of a limited liability corporation (LLC)?

I. limited liability for all members

II. no taxation as an entity (no double taxation)

III. more than one class of stock

IV. limited to fewer than 25 members

V. fairly easy to establish

VI. no "continuity of life" like regular corporation

(A) I, II, and IV

(B) I, II, III, and VI

(C) I, II, III, IV, and V

(D) I, III, IV, V, and VI

12. Which of the following statements is FALSE in regard to joint ventures?

(A) Members of a joint venture may be any combination of sole proprietorships, partnerships, and corporations.

(B) A joint venture is a business entity separate from its members.

(C) A joint venture spreads risk and rewards, and it pools expertise, experience, and resources. However, bonding capacity is not aggregated.

(D) A joint venture usually dissolves after the completion of a specific project.

13. Which of the following construction business types can have unlimited shareholders?

I.   S corporation

II.   LLC

III. corporation

IV.   sole proprietorship

(A) II and III

(B) I, II, and III

(C) I, III, and IV

(D) I, II, III, and IV

14. The phrase "or approved equal" allows a contractor to

(A) substitute one connection design for another

(B) substitute a more expensive feature for another

(C) replace an open-shop subcontractor with a union subcontractor

(D) install a product whose brand name and model number are not listed in the specifications

15. Cities, other municipalities, and departments of transportation often have standard specifications, in addition to the specifications issued as part of the construction document set, that cover such items as

(A) safety requirements

(B) environmental requirements

(C) concrete, fire hydrant, manhole structures, and curb requirements

(D) procurement and accounting requirements

16. What is intended to prevent a contractor from bidding on a project and subsequently backing out after being selected for the project?

(A) publically recorded bid

(B) property lien

(C) surety bond

(D) proposal bond

17. Which of the following is illegal, in addition to being unethical?

(A) bid shopping

(B) bid peddling

(C) bid rigging

(D) bid unbalancing

18. Which of the following is NOT normally part of a construction contract?

I.   invitation to bid

II.   instructions to bidders

III.   general conditions

IV.   supplementary conditions

V.   liability insurance policy

VI.   technical specifications

VII.   drawings

VIII. addenda

IX.   proposals

X.   bid bond

XI.   agreement

XII.   performance bond

XIII. labor and material payment bond

XIV.   nondisclosure agreement

(A) I

(B) II

(C) V

(D) XIV

19. Once a contract has been signed by the owner and contractor, changes to the contract

(A) cannot be made

(B) can be made by the owner, but not by the contractor

(C) can be made by the contractor, but not by the owner

(D) can be made by both the owner and the contractor

20. A constructive change is a change to the contract that can legally be construed to have been made, even though the owner did not issue a specific, written change order. Which of the following situations is normally a constructive change?

(A) request by the engineer-architect to install OSHA-compliant safety features

(B) delay caused by the owner's failure to provide access

(C) rework mandated by the building official

(D) expense and delay due to adverse weather

21. A bid for foundation construction is based on owner supplied soil borings showing sandy clay to a depth of 12 ft. However, after the contract has been assigned and during construction, the backhoe encounters large pieces of concrete buried throughout the construction site. This situation would normally be referred to as

(A) concealed conditions

(B) unexplained features

(C) unexpected characteristics

(D) hidden detriment

22. If a contract has a value engineering clause and a contractor suggests to the owner that a feature or method be used to reduce the annual maintenance cost of the finished project, what will be the most likely outcome?

(A) The contractor will be able to share one time in the owner's expected cost savings.

(B) The contractor will be paid a fixed amount (specified by the contract) for making suggestion, but only if the suggestion is accepted.

(C) The contract amount will be increased by some amount specified in the contract.

(D) The contractor will receive an annuity payment over some time period specified in the contract.

23. A contract has a value engineering clause that allows the parties to share in improvements that reduce cost. The contractor had originally planned to transport concrete on site for a small pour with motorized wheelbarrows. On the day of the pour, however, a concrete pump is available and is used, substantially reducing the contractor's labor cost for the day. This is an example of

(A) value engineering whose benefit will be shared by both contractor and owner

(B) efficient methodology whose benefit is to the contractor only

(C) value engineering whose benefit is to the owner only

(D) cost reduction whose benefit will be shared by both contractor and laborers

24. A material breach of contract occurs when

(A) the contractor uses material not approved by the contract to use

(B) the contractor's material order arrives late

(C) the owner becomes insolvent

(D) the contractor installs a feature incorrectly

Professional

SOLUTIONS

1. The three different forms of company ownership are the (1) sole proprietorship, (2) partnership, and (3) corporation.

A *sole proprietor* is his or her own boss. This satisfies the proprietor's ego and facilitates quick decisions, but unless the proprietor is trained in business, the company will usually operate without the benefit of expert or mitigating advice. The sole proprietor also personally assumes all the debts and liabilities of the company. A sole proprietorship is terminated upon the death of the proprietor.

A *partnership* increases the capitalization and the knowledge base beyond that of a proprietorship, but offers little else in the way of improvement. In fact, the partnership creates an additional disadvantage of one partner's possible irresponsible actions creating debts and liabilities for the remaining partners.

A *corporation* has sizable capitalization (provided by the stockholders) and a vast knowledge base (provided by the board of directors). It keeps the company and owner liability separate. It also survives the death of any employee, officer, or director. Its major disadvantage is the administrative work required to establish and maintain the corporate structure.

2. To be legal, a contract must contain an *offer*, some form of *consideration* (which does not have to be equitable), and an *acceptance* by both parties. To be enforceable, the contract must be voluntarily entered into, both parties must be competent and of legal age, and the contract cannot be for illegal activities.

3. A written contract will identify both parties, state the purpose of the contract and the obligations of the parties, give specific details of the obligations (including relevant dates and deadlines), specify the consideration, state the boilerplate clauses to clarify the contract terms, and leave places for signatures.

4. A consultant will either charge a fixed fee, a variable fee, or some combination of the two. A one-time fixed fee is known as a *lump-sum fee*. In a *cost plus fixed fee* contract, the consultant will also pass on certain costs to the client. Some charges to the client may depend on other factors, such as the salary of the consultant's staff, the number of days the consultant works, or the eventual cost or value of an item being designed by the consultant.

5. A *retainer* is a (usually) nonreturnable advance paid by the client to the consultant. While the retainer may be intended to cover the consultant's initial expenses until the first big billing is sent out, there does not need to be any rational basis for the retainer. Often, a small retainer is used by the consultant to qualify the client

(i.e., to make sure the client is not just shopping around and getting free initial consultations) and as a security deposit (to make sure the client does not change consultants after work begins).

6. The Engineers Joint Contract Documents Committee (EJCDC) consists of the National Society of Professional Engineers, the American Council of Engineering Companies (formerly the American Consulting Engineers Council), the American Society of Civil Engineers, Construction Specifications Institute, and the Associated General Contractors of America. The American Institute of Architects is not a member, and it has its own standardized contract documents.

The answer is (D).

7. A business in the construction industry must have two or more employees and a minimum annual gross sales volume of $500,000 to be subject to the Fair Labor Standards Act (FLSA). Individual coverage also applies to employees whose work regularly involves them in commerce between states (i.e., interstate commerce). Any person who works on, or otherwise handles, goods moving in interstate commerce, or who works on the expansion of existing facilities of commerce, is individually subject to the protection of the FLSA and the current minimum wage and overtime pay requirements, regardless of the sales volume of the employer.

The answer is (B).

8. A *double-breasted* design firm serves both union and nonunion clients. When union-affiliated companies find themselves uncompetitive in bidding on nonunion projects, the company owners may decide to form and operate a second company that is *open shop* (i.e., employees are not required to join a union as a condition of employment). Although there are some restrictions requiring independence of operation, the common ownership of two related firms is legal.

The answer is (C).

9. *Without expressed authority* means each member of a partnership has full authority to obligate the partnership (and the other partners).

The answer is (A).

10. Only general partners can vote in a partnership. Both silent and limited partners share in the profit and benefits of the partnership, but they only contribute financing and do not participate in the management. The identities of silent partners are often known only to a few, whereas limited partners are known to all.

The answer is (B).

Professional

11. Limited liability corporations have limited liability for all members, no double taxation, more than one type of stock, and no continuity of life. They are not limited in members, and they are comparatively fairly difficult to establish.

The answer is (B).

12. One of the reasons for forming joint ventures is that the bonding capacity is aggregated. Even if each contractor cannot individually meet the minimum bond requirements, the total of the bonding capacities may be sufficient.

The answer is (C).

13. Both normal corporations and limited liability corporations (LLCs) can have unlimited shareholders. S corporations and sole proprietorships are for individuals.

The answer is (A).

14. When the specifications include a nonstructural, brand-named article and the accompanying phrase "or approved equal," the contractor can substitute something with the same functionality, even though it is not the brand-named article. "Or approved equal" would not be used with a structural detail such as a connection design.

The answer is (D).

15. Municipalities that experience frequent construction projects within their boundaries have standard specifications that are included by reference in every project's construction document set. This document set would cover items such as concrete, fire hydrants, manhole structures, and curb requirements.

The answer is (C).

16. *Proposal bonds*, also known as *bid bonds*, are insurance policies payable to the owner in the event that the contractor backs out after submitting a qualified bid.

The answer is (D).

17. *Bid rigging*, also known as *price fixing*, is an illegal arrangement between contractors to control the bid prices of a construction project or to divide up customers or market areas. Bid shopping before or after the bid letting is where the general contractor tries to secure better subcontract proposals by negotiating with the subcontractors. Bid peddling is done by the subcontractor to try to lower its proposal below the lowest proposal. Bid unbalancing is where a contractor pushes the payment for some expense items to prior construction phases in order to improve cash flow.

The answer is (C).

18. The construction contract includes many items, some explicit only by reference. The contractor may be required to carry liability insurance, but the policy itself is between the contractor and its insurance company, and is not normally part of the construction contract.

The answer is (C).

19. Changes to the contract can be made by either the owner or the contractor. The method for making such changes is indicated in the contract. Almost always, the change must be agreed to by both parties in writing.

The answer is (D).

20. A constructive change to the contract is the result of an action or lack of action of the owner or its agent. If the project is delayed by the owner's failure to provide access, the owner has effectively changed the contract.

The answer is (B).

21. *Concealed conditions* are also known as *changed conditions* and *differing site conditions*. Most, but not all, contracts have provisions dealing with changed conditions. Some place the responsibility to confirm the site conditions before bidding on the contractor. Others detail the extent of changed conditions that will trigger a review of reimbursable expenses.

The answer is (A).

22. Changes to a structure's performance, safety, appearance, or maintenance that benefit the owner in the long run will be covered by the value engineering clause of a contract. Normally, the contractor is able to share in cost savings in some manner by receiving a payment or credit to the contract.

The answer is (A).

23. The problem gives an example of efficient methodology, where the benefit is to the contractor only. It is not an example of value engineering, as the change affects the contractor, not the owner. Performance, safety, appearance, and maintenance are unaffected.

The answer is (B).

24. *A material breach of the contract* is a significant event that is grounds for cancelling the contract entirely. Typical triggering events include failure of the owner to make payments, the owner causing delays, the owner declaring bankruptcy, the contractor abandoning the job, or the contractor getting substantially off schedule.

The answer is (C).

Professional

58 Engineering Ethics

PRACTICE PROBLEMS

(Each problem has two parts. Determine whether the situation is (or can be) permitted legally. Then, determine whether the situation is permitted ethically.)

1. (a) Was it legal and/or ethical for an engineer to sign and seal plans that were not prepared by him or prepared under his responsible direction, supervision, or control?

(b) Was it legal and/or ethical for an engineer to sign and seal plans that were not prepared by him but were prepared under his responsible direction, supervision, and control?

2. Under what conditions would it be legal and/or ethical for an engineer to rely on the information (e.g., elevations and amounts of cuts and fills) furnished by a grading contractor?

3. Was it legal and/or ethical for an engineer to alter the soils report prepared by another engineer for his client?

4. Under what conditions would it be legal and/or ethical for an engineer to assign work called for in his contract to another engineer?

5. A licensed professional engineer was convicted of a felony totally unrelated to his consulting engineering practice.

(a) What actions would you recommend be taken by the state registration board?

(b) What actions would you recommend be taken by the professional or technical society (e.g., ASCE, ASME, IEEE, NSPE, and so on)?

6. An engineer came across some work of a predecessor. After verifying the validity and correctness of all assumptions and calculations, the engineer used the work. Under what conditions would such use be legal and/or ethical?

7. A building contractor made it a policy to provide cell phones to the engineers of projects he was working on. Under what conditions could the engineers accept the phones?

8. An engineer designed a tilt-up slab building for a client. The design engineer sent the design out to another engineer for checking. The checking engineer sent the plans to a concrete contractor for review. The concrete contractor made suggestions that were incorporated into the recommendations of the checking contractor. These recommendations were subsequently incorporated into the plans by the original design engineer. What steps must be taken to keep the design process legal and/or ethical?

9. A consulting engineer registered his corporation as "John Williams, P.E. and Associates, Inc." even though he had no associates. Under what conditions would this name be legal and/or ethical?

10. When it became known that a chemical plant was planning on producing a toxic product, an engineer wrote to the local newspaper condemning the chemical plant's action. Under what conditions would the engineer's action be legal and/or ethical?

11. An engineer signed a contract with a client. The fee the client agreed to pay was based on the engineer's estimate of time required. The engineer was able to complete the contract satisfactorily in half the time he expected. Under what conditions would it be legal and/or ethical for the engineer to keep the full fee?

12. After working on a project for a client, the engineer was asked by a competitor of the client to perform design services. Under what conditions would it be legal and/or ethical for the engineer to work for the competitor?

13. Two engineers submitted bids to a prospective client for a design project. The client told engineer A how much engineer B had bid and invited engineer A to beat the amount. Under what conditions could engineer A legally/ethically submit a lower bid?

14. A registered civil engineer specializing in well-drilling, irrigation pipelines, and farmhouse sanitary systems took a booth at a county fair located in a farming town. By a random drawing, the engineer's booth was located next to a hog-breeder's booth, complete with live (prize) hogs. The engineer gave away helium balloons with his name and phone number to all visitors to the booth. Did the engineer violate any laws/ethical guidelines?

15. While in a developing country supervising construction of a project an engineer designed, the engineer discovered the client's project manager was treating local workers in an unsafe and inhumane (but for that country, legal) manner. When the engineer objected, the client told the engineer to mind his own business. Later, the local workers asked the engineer to participate in a walkout and strike with them.

(a) What legal/ethical positions should the engineer take?

(b) Should it have made any difference if the engineer had or had not yet accepted any money from the client?

16. While working for a client, an engineer learns confidential knowledge of a proprietary production process being used by the client's chemical plant. The process is clearly destructive to the environment, but the client will not listen to the objections of the engineer. To inform the proper authorities will require the engineer to release information that was gained in confidence. Is it legal and/or ethical for the engineer to expose the client?

17. While working for an engineering design firm, an engineer was moonlighting as a soils engineer. At night, the engineer used the employer's facilities to analyze and plot the results of soils tests. He then used his employer's computers and word processors to write his reports. The equipment, computers, and word processors would otherwise be unused. Under what conditions could the engineer's actions be considered legal and/or ethical?

18. Ethical codes and state legislation forbidding competitive bidding by design engineers are

(A) enforceable in some states

(B) not enforceable on public (nonfederal) projects

(C) enforceable for projects costing less than $5 million dollars

(D) not enforceable

SOLUTIONS

Introduction to the Solutions

Case studies in law and ethics can be interpreted in many ways. The problems presented are simple thumbnail outlines. In most real cases, there will be more facts to influence a determination than are presented in the case scenarios. In some cases, a state may have specific laws affecting the determination; in other cases, prior case law will have been established.

The determination of whether an action is legal can be made in two ways. The obvious interpretation of an illegal action is one that violates a specific law or statute. An action can also be *found to be illegal* if it is judged in court to be a breach of a written, verbal, or implied contract. Both of these approaches are used in the following solutions.

These answers have been developed to teach legal and ethical principals. While being realistic, they are not necessarily based on actual incidents or prior case law.

1. (a) Stamping plans for someone else is illegal. The registration laws of all states permit a registered engineer to stamp/sign/seal only plans that were prepared by him personally or were prepared under his direction, supervision, or control. This is sometimes called being in *responsible charge*. The stamping/signing/sealing, for a fee or gratis, of plans produced by another person, whether that person is registered or not and whether that person is an engineer or not, is illegal.

(b) The act is unethical. An illegal act, being a concealed act, is intrinsically unethical. In addition, stamping/signing/sealing plans that have not been checked violates the rule contained in all ethical codes that requires an engineer to protect the public.

2. Unless the engineer and contractor worked together such that the engineer had personal knowledge that the information was correct, accepting the contractor's information is illegal. Not only would using unverified data violate the state's registration law (for the same reason that stamping/signing/sealing unverified plans in Prob. 1 was illegal), but the engineer's contract clause dealing with assignment of work to others would probably be violated.

The act is unethical. An illegal act, being a concealed act, is intrinsically unethical. In addition, using unverified data violates the rule contained in all ethical codes that requires an engineer to protect the client.

3. It is illegal to alter a report to bring it "more into line" with what the client wants unless the alterations represent actual, verified changed conditions. Even when the alterations are warranted, however, use of the unverified remainder of the report is a violation of the state registration law requiring an engineer only to stamp/sign/seal plans developed by or under him.

Furthermore, this would be a case of fraudulent misrepresentation unless the originating engineer's name was removed from the report.

Unless the engineer who wrote the original report has given permission for the modification, altering the report would be unethical.

4. Assignment of engineering work is legal (1) if the engineer's contract permitted assignment, (2) all prerequisites (i.e., notifying the client) were met, and (3) the work was performed under the direction of another licensed engineer.

Assignment of work is ethical (1) if it is not illegal, (2) if it is done with the awareness of the client, and (3) if the assignor has determined that the assignee is competent in the area of the assignment.

5. (a) The registration laws of many states require a hearing to be held when a licensee is found guilty of unrelated, but nevertheless unforgivable, felonies (e.g., moral turpitude). The specific action (e.g., suspension, revocation of license, public censure, and so on) taken depends on the customs of the state's registration board.

(b) By convention, it is not the responsibility of technical and professional organizations to monitor or judge the personal actions of their members. Such organizations do not have the authority to discipline members (other than to revoke membership), nor are they immune from possible retaliatory libel/slander lawsuits if they publicly censure a member.

6. The action is legal because, by verifying all the assumptions and checking all the calculations, the engineer effectively does the work. Very few engineering procedures are truly original; the fact that someone else's effort guided the analysis does not make the action illegal.

The action is probably ethical, particularly if the client and the predecessor are aware of what has happened (although it is not necessary for the predecessor to be told). It is unclear to what extent (if at all) the predecessor should be credited. There could be other extenuating circumstances that would make referring to the original work unethical.

7. Gifts, per se, are not illegal. Unless accepting the phones violates some public policy or other law, or is in some way an illegal bribe to induce the engineer to favor the contractor, it is probably legal to accept the phones.

Ethical acceptance of the phones requires (among other considerations) that (1) the phones be required for the job, (2) the phones be used for business only, (3) the phones are returned to the contractor at the end of the

job, and (4) the contractor's and engineer's clients know and approve of the transaction.

8. There are two issues: (1) the assignment and (2) the incorporation of work done by another. To avoid a breach, the contracts of both the design and checking engineers must permit the assignments. To avoid a violation of the state registration law requiring engineers to be in responsible charge of the work they stamp/sign/seal, both the design and checking engineers must verify the validity of the changes.

To be ethical, the actions must be legal and all parties (including the design engineer's client) must be aware that the assignments have occurred and that the changes have been made.

9. The name is probably legal. If the name was accepted by the state's corporation registrar, it is a legally formatted name. However, some states have engineering registration laws that restrict what an engineering corporation may be named. For example, all individuals listed in the name (e.g., "Cooper, Williams, and Somerset—Consulting Engineers") may need to be registered. Whether having "Associates" in the name is legal depends on the state.

Using the name is unethical. It misleads the public and represents unfair competition with other engineers running one-person offices.

10. Unless the engineer's accusation is known to be false or exaggerated, or the engineer has signed an agreement (e.g., confidentiality, non-disclosure, etc.) with his employer forbidding the disclosure, the letter to the newspaper is probably not illegal.

The action is probably unethical. (If the letter to the newspaper is unsigned it is a concealed action and is definitely unethical.) While whistle-blowing to protect the public is implicitly an ethical procedure, unless the engineer is reasonably certain that manufacture of the toxic product represents a hazard to the public, he has a responsibility to the employer. Even then, the engineer should exhaust all possible remedies to render the manufacture nonhazardous before blowing the whistle. Of course, the engineer may quit working for the chemical plant and be as critical as the law allows without violating engineer-employer ethical considerations.

11. Unless the engineer's payment was explicitly linked in the contract to the amount of time spent on the job, taking the full fee would not be illegal or a breach of the contract.

An engineer has an obligation to be fair in estimates of cost, particularly when the engineer knows no one else is providing a competitive bid. Taking the full fee would be ethical if the original estimate was arrived at logically and was not meant to deceive or take advantage of the client. An engineer is permitted to take advantage of economies of scale, state-of-the-art techniques, and break-through methods. (Similarly, when a job costs more than the estimate, the engineer may be ethically bound to stick with the original estimate.)

12. In the absence of a nondisclosure or noncompetition agreement or similar contract clause, working for the competitor is probably legal.

Working for both clients is unethical. Even if both clients know and approve, it is difficult for the engineer not to "cross-pollinate" his work and improve one client's position with knowledge and insights gained at the expense of the other client. Furthermore, the mere appearance of a conflict of interest of this type is a violation of most ethical codes.

13. In the absence of a sealed-bid provision mandated by a public agency and requiring all bids to be opened at once (and the award going to the lowest bidder), the action is probably legal.

It is unethical for an engineer to undercut the price of another engineer. Not only does this violate a standard of behavior expected of professionals, it unfairly benefits one engineer because a similar chance is not given to the other engineer. Even if both engineers are bidding openly against each other (in an auction format), the client must understand that a lower price means reduced service. Each reduction in price is an incentive to the engineer to reduce the quality or quantity of service.

14. It is generally legal for an engineer to advertise his services. Unless the state has relevant laws, the engineer probably did not engage in illegal actions.

Most ethical codes prohibit unprofessional advertising. The unfortunate location due to a random drawing might be excusable, but the engineer should probably refuse to participate. In any case, the balloons are a form of unprofessional advertising, and as such, are unethical.

15. (a) As stated in the scenario statement, the client's actions are legal for that country. The fact that the actions might be illegal in another country is irrelevant. Whether or not the strike is legal depends on the industry and the laws of the land. Some or all occupations (e.g., police and medical personnel) may be forbidden to strike. Assuming the engineer's contract does not prohibit participation, the engineer should determine the legality of the strike before making a decision to participate.

If the client's actions are inhuman, the engineer has an ethical obligation to withdraw from the project. Not doing so associates the profession of engineering with human misery.

(b) The engineer has a contract to complete the project for the client. (It is assumed that the contract between the engineer and client was negotiated in good faith, that the engineer had no knowledge of the work conditions prior to signing, and that the client did not falsely induce the engineer to sign.) Regardless of the reason for withdrawing, the engineer is breaching his contract. In the absence of proof of illegal actions by the client, withdrawal by the engineer requires a return of all fees received. Even if no fees have been received, withdrawal exposes the engineer to other delay-related claims by the client.

16. A contract for an illegal action cannot be enforced. Therefore, any confidentiality or nondisclosure agreement that the engineer has signed is unenforceable if the production process is illegal, uses illegal chemicals, or violates laws protecting the environment. If the production process is not illegal, it is not legal for the engineer to expose the client.

Society and the public are at the top of the hierarchy of an engineer's responsibilities. Obligations to the public take precedence over the client. If the production process is illegal, it would be ethical to expose the client.

17. It is probably legal for the engineer to use the facilities, particularly if the employer is aware of the use. (The question of whether the engineer is trespassing or violating a company policy cannot be answered without additional information.)

Moonlighting, in general, is not ethical. Most ethical codes prohibit running an engineering consulting business while receiving a salary from another employer. The rationale is that the moonlighting engineer is able to offer services at a much lower price, placing other consulting engineers at a competitive disadvantage. The use of someone else's equipment compounds the problem since the engineer does not have to pay for using the equipment, and so does not have to charge any clients for it. This places the engineer at an unfair competitive advantage compared to other consultants who have invested heavily in equipment.

18. Ethical bans on competitive bidding are not enforceable. The National Society of Professional Engineers' (NSPE) ethical ban on competitive bidding was struck down by the U.S. Supreme Court in 1978 as a violation of the Sherman Antitrust Act of 1890.

The answer is (D).

59 Electrical Engineering Frontiers

PRACTICE PROBLEMS

1. What are the force particles responsible for the electromagnetic interaction?

(A) bosons

(B) gluons

(C) gravitons

(D) photons

SOLUTIONS

1. The force particles for electromagnetic interactions are photons, sometimes called virtual photons because they cannot be detected without interfering with the interaction.

The answer is (D).

60 Engineering Licensing in the United States

PRACTICE PROBLEMS

1. What might be a reason to obtain a PE license?

(A) increased respect

(B) increased wages

(C) proof of one's engineering knowledge

(D) all of the above

SOLUTIONS

1. Obtaining a PE license is proof to peers of the broad base of your knowledge, and an example of your self discipline. The PE will engender greater respect and is known to have a positive impact on salary.

The answer is (D).

WITHDRAWAL